FINLAND

Helsinki
olm
Tallinn ESTONIA
Riga LATVIA
Vilnius LITHUANIA · Moscow

AND
BELARUS Minsk

R U S S I A

· Nur-Sultan

LOVAKIA UKRAINE
HUNGARY MOLDOVA
Chisinau
SERBIA ROMANIA
Belgrade · Bucharest
KOS Sofia
Pristina BULGARIA
LBANIA
hens
GREECE GEORGIA Tbilisi
ARMENIA AZERBAIJAN
Ankara Yerevan Baku
CYPRUS Nicosia
LEBANON Beirut
ISRAEL Damascus
Jerusalem SYRIA
Amman
Cairo JORDAN

K A Z A K H S T A N

UZBEKISTAN
Tashkent Bishkek
KYRGYZSTAN
TURKMENISTAN TAJIKISTAN
Ashgabat Dushanbe

M O N G O L I A
· Ulaanbaatar

NORTH
KOREA
· Pyongyang JAPAN
· Seoul
SOUTH · Sejong City
KOREA · Tokyo
· Beijing

TURKEY

IRAQ I R A N
Baghdad
KUWAIT
Kuwait
BAHRAIN
Manama QATAR
Riyadh · Doha
SAUDI UAE · Muscat
ARABIA Abu Dhabi

EGYPT

Tehran

Kabul
AFGHANISTAN
Islamabad

PAKISTAN

C H I N A

Taipei
TAIWAN

New Delhi NEPAL
Thimphu
Kathmandu BHUTAN
BANGLADESH
Dhaka MYANMAR Hanoi
(BURMA)
Nay Pyi Taw VIETNAM

SUDAN
Khartoum

YEMEN OMAN
ERITREA
Asmara
Sanaa
DJIBOUTI
Djibouti

SOMALIA

I N D I A

Socotra
(Yemen)

Laccadive Islands
(India)
Colombo
Sri Jayewardenepura Kotte
SRI LANKA

Andaman
Islands
(India)

LAOS
Vientiane
THAILAND
Bangkok
CAMBODIA
Phnom Penh

Manila

PHILIPPINES

Northern
Mariana
Islands
(US)

· Guam
(US)

MARSHALL ISLANDS
Majuro Atoll

· Ngerulmud
Palikir

NTRAL
N REPUBLIC
angui

SOUTH
SUDAN
· Juba

ETHIOPIA
Addis Ababa

Mogadishu

UGANDA
Kampala
RWANDA KENYA
Kigali Nairobi
DEM. REP. BURUNDI
CONGO Bujumbura Dodoma
TANZANIA

Nicobar
Islands
(India)
MALDIVES
Male'

Kuala Lumpur MALAYSIA
Putrajaya
SINGAPORE
Singapore

BRUNEI
Bandar Seri Begawan

PALAU MICRONESIA

NAURU
Yaren

· Tarawa Atoll

K I R I B A T I

SEYCHELLES
· Victoria

British Indian
Ocean Territory
(UK)

I N D O N E S I A

Jakarta

PAPUA NEW GUINEA

SOLOMON
ISLANDS
Honiara

Funafuti Atoll

TUVALU

Tokelau
(NZ)

ZAMBIA MALAWI
Lusaka
MOZAMBIQUE
COMOROS
Moroni
Mayotte
(France)

Christmas Island
(Australia)

Dili EAST
TIMOR

Port Moresby

Coral Sea
Islands
(Australia)

VANUATU
· Port-Vila
FIJI

New
Caledonia
(France)

Wallis
& Futuna
(France) SAMOA
Apia

Suva

TONGA
Nuku'alofa

A
ZIMBABWE
Harare

BOTSWANA
Gaborone
Pretoria
emfontein Maputo
Mbabane
Maseru ESWATINI (SWAZILAND)
SOUTH
AFRICA

Antananarivo

MADAGASCAR

MAURITIUS
Port Louis

Réunion
(France)

WESTERN
AUSTRALIA

NORTHERN
TERRITORY

QUEENSLAND

SOUTH
AUSTRALIA

A U S T R A L I A
NEW SOUTH
WALES

· Canberra
VICTORIA AUSTRALIAN
CAPITAL
TERRITORY

NEW ZEALAND
· Wellington

TASMANIA

Prince Edward
Islands
(South Africa)

Crozet Islands
(France)

Kerguelen
(France)

Chatham Islands
(New Zealand)

Auckland Islands
(New Zealand)

Macquarie Island
(Australia)

Country abbreviations

BEL.	Belgium
BOS. & HERZ.	Bosnia and Herzegovina
KOS.	Kosovo (disputed)
LIECH.	Liechtenstein
LUX.	Luxembourg
N. MAC.	North Macedonia
MON.	Montenegro
NETH.	Netherlands
NZ	New Zealand
SM	San Marino
SLVN.	Slovenia
SWITZ.	Switzerland
UAE	United Arab Emirates
UK	United Kingdom
US	United States of America
VAT. CITY	Vatican City

A N T A R C T I C A

CLIMATE EMERGENCY ATLAS

CLIMATE EMERGENCY ATLAS

WRITER AND RESEARCHER DAN HOOKE

CONSULTANTS PROFESSOR FRANS BERKHOUT

AND PROFESSOR KIRSTIN DOW

DK Penguin Random House

Project Editor Sam Kennedy
Senior Art Editor Rachael Grady
Senior Cartographic Editor Simon Mumford
Senior Contributing Editors Georgina Palffy, Jenny Sich,
Anna Streiffert-Limerick, Selina Wood
Editor Kelsie Besaw
Designers Kit Lane, Mik Gates, Lynne Moulding, Greg McCarthy
Illustrators Jon @ KJA Artists, Adam Benton
Managing Editor Francesca Baines
Managing Art Editor Philip Letsu
Production Editor Robert Dunn
Production Controller Jude Crozier
Jacket Designer Akiko Kato
Design Development Manager Sophia MTT
Picture Research Geetika Bhandari, Surya Sarangi

First published in Great Britain in 2020 by
Dorling Kindersley Limited
DK, One Embassy Gardens, 8 Viaduct Gardens,
London, SW11 7BW

A CIP catalogue record for this book
is available from the British Library.
ISBN: 978-0-2414-4643-0

Printed and bound in the UAE

For the curious
www.dk.com

MIX
Paper from
responsible sources
FSC™ C018179

This book was made with Forest Stewardship Council® certified
paper – one small step in DK's commitment to a sustainable future.
For more information go to
www.dk.com/our-green-pledge

Measuring greenhouse gases
When measuring the impact that a combination of gases has on warming the atmosphere, this book uses the unit Greenhouse Gases (GHGs). This unit combines multiple gases by comparing their warming potential to that of carbon dioxide.

Coronavirus 2020
This book was made in 2020, during the Coronavirus outbreak. All information is accurate at the time of going to press, but it is too early to know how the pandemic will affect future climate change policies.

CONTENTS

How Earth's climate works

The causes of climate change

Heatwaves

Bushfires

Drylands

Foreword

By Liz Bonnin, science, natural history and environmental broadcaster.

Our planet is an awe-inspiring place. From icy mountain habitats to the mysterious deep sea, its spectacular ecosystems are home to an extraordinary array of species. Each has evolved over millions of years to live in perfect balance with others and carry out its essential role in keeping the Earth healthy – and when the Earth thrives, we thrive.

But the way we live is harming our planet. We're using up its natural resources to produce fuels, fast fashion, and all sorts of gadgets that we replace constantly, producing so many greenhouse gases that the world is heating up. On my travels to discover how wildlife is coping, I've seen vast stretches of bleached corals, stripped of all the colourful life they support because of rising sea temperatures, and huge patches of rainforest, crucial for absorbing CO_2 and stabilizing our climate, cleared away and

replaced with cattle that overgraze the grass until nothing remains.

But I've also had the privilege of meeting my heroes – the scientists who are working day and night to save our wild places. Thanks to their dedication, some incredible things are happening: researchers are working to recolonize the Great Barrier Reef, attracting baby fish looking for new homes by playing the sounds of healthy reefs through underwater speakers. And high in the canopy of the Amazon, camera traps have been fitted to harpy eagle nests, to help protect the chicks and their precious rainforest.

Each of us needs to play our part too, and this book is full of information about why this is so important. Eating less meat, using green energy in our homes, and writing to our MPs demanding change sends out a clear message that we want to live differently, and it inspires others around us to do the same. Our voices are far louder than we might first imagine! We're in this together, we can be the change the planet needs.

Liz Bonnin

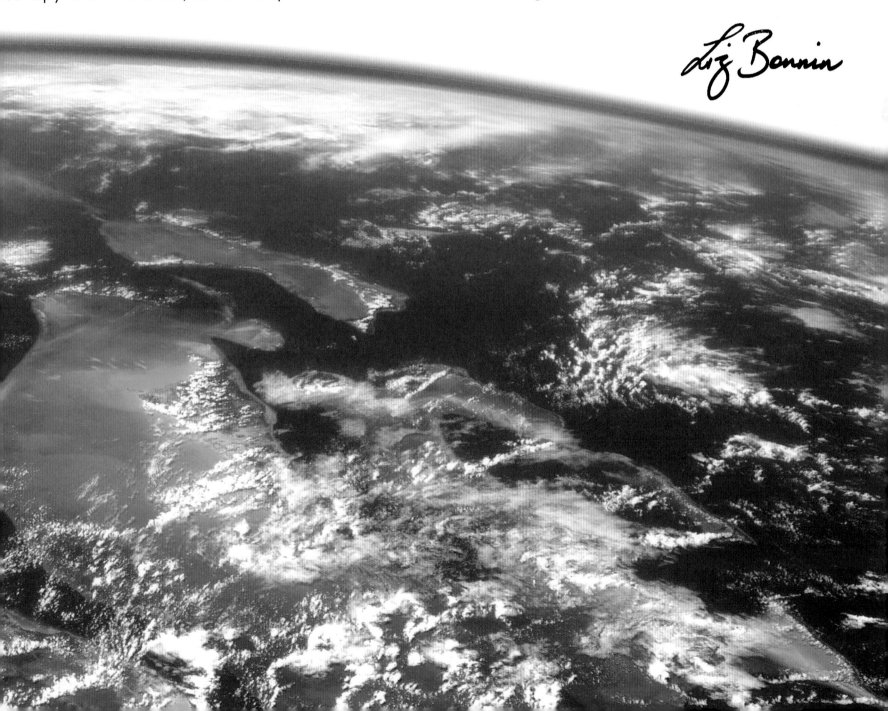

How Earth's climate works

What is the climate emergency?

All around the world, human activity is causing the emission of greenhouse gases – so called because they trap heat on Earth. This is known as global warming and it is affecting both the climate and life on our planet. To have any chance of halting the rise in temperature, and even more dramatic effects in the future, we need to act now.

Human activity

Human activity – such as burning fossil fuels, deforestation, and cattle farming – emits greenhouse gases.

Greenhouse effect

Greenhouse gases build up in the atmosphere, trapping more and more heat.

Weather or climate?
The difference between climate and weather is that weather is the short-term conditions that you experience day to day – such as sun or rain. Climate is the average weather in a region over a longer period, typically 30 years. While the clothes you pick to wear each day reflect the weather, the variety of clothes in your wardrobe depends on the climate.

Weather

Climate

Ice melt

Melting ice from Greenland, Antarctica, and glaciers on land contributes to sea-level rise, while Arctic sea ice has shrunk dramatically.

Ocean damage

The oceans are warming and rising, endangering coastal communities and marine wildlife.

Temperature rise

The direct impact of the greenhouse effect is an increase in temperature, or "global warming". This creates knock-on effects across the world's climate system.

Homes at risk

People's homes and livelihoods are threatened by sea-level rise, droughts, and wildfires.

Habitat loss

Animal habitats are changing and being destroyed by changes in the climate, risking the extinction of some species.

Extreme weather

Climate change has increased the frequency of extreme weather events, such as cyclones and heatwaves.

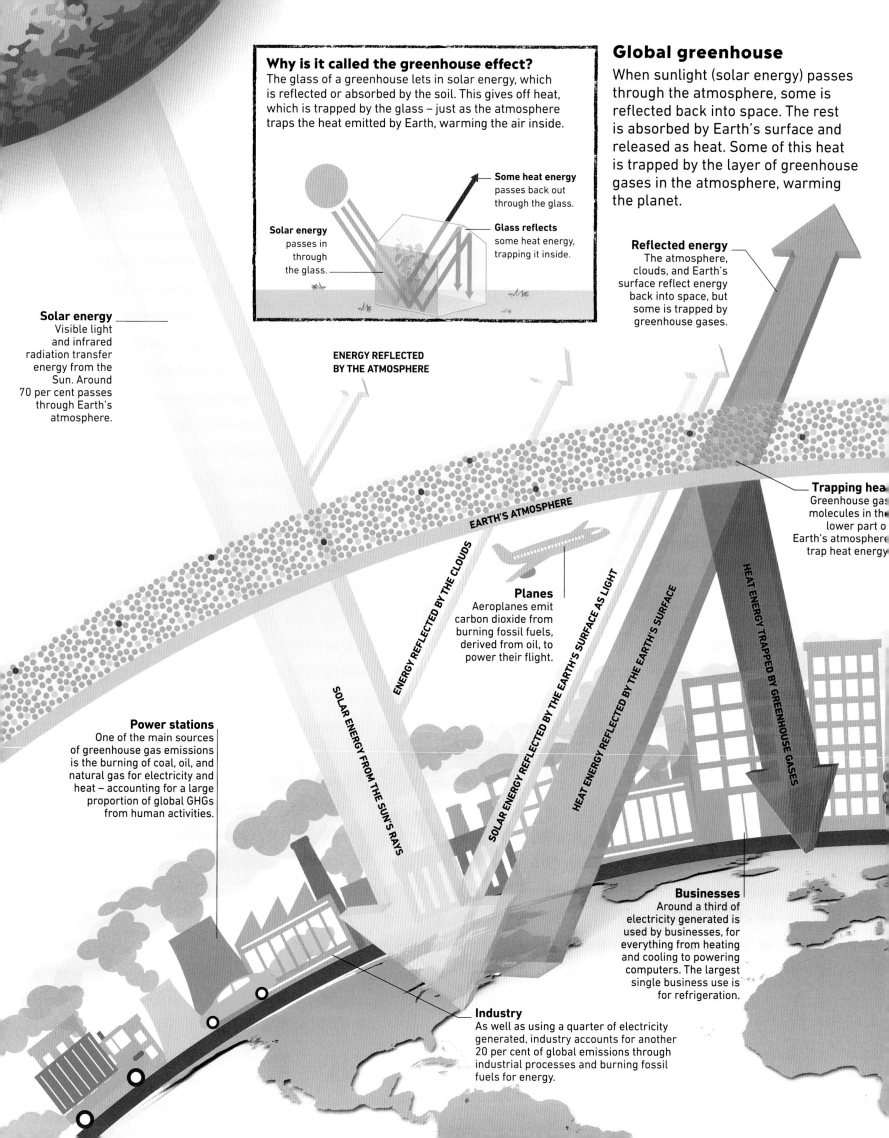

Why is it called the greenhouse effect?

The glass of a greenhouse lets in solar energy, which is reflected or absorbed by the soil. This gives off heat, which is trapped by the glass – just as the atmosphere traps the heat emitted by Earth, warming the air inside.

Some heat energy passes back out through the glass.

Solar energy passes in through the glass.

Glass reflects some heat energy, trapping it inside.

Global greenhouse

When sunlight (solar energy) passes through the atmosphere, some is reflected back into space. The rest is absorbed by Earth's surface and released as heat. Some of this heat is trapped by the layer of greenhouse gases in the atmosphere, warming the planet.

Reflected energy
The atmosphere, clouds, and Earth's surface reflect energy back into space, but some is trapped by greenhouse gases.

ENERGY REFLECTED BY THE ATMOSPHERE

Solar energy
Visible light and infrared radiation transfer energy from the Sun. Around 70 per cent passes through Earth's atmosphere.

EARTH'S ATMOSPHERE

ENERGY REFLECTED BY THE CLOUDS

Trapping hea
Greenhouse gas molecules in the lower part o Earth's atmosphere trap heat energy

Planes
Aeroplanes emit carbon dioxide from burning fossil fuels, derived from oil, to power their flight.

SOLAR ENERGY FROM THE SUN'S RAYS

SOLAR ENERGY REFLECTED BY THE EARTH'S SURFACE AS LIGHT

HEAT ENERGY REFLECTED BY THE EARTH'S SURFACE

HEAT ENERGY TRAPPED BY GREENHOUSE GASES

Power stations
One of the main sources of greenhouse gas emissions is the burning of coal, oil, and natural gas for electricity and heat – accounting for a large proportion of global GHGs from human activities.

Businesses
Around a third of electricity generated is used by businesses, for everything from heating and cooling to powering computers. The largest single business use is for refrigeration.

Industry
As well as using a quarter of electricity generated, industry accounts for another 20 per cent of global emissions through industrial processes and burning fossil fuels for energy.

The greenhouse **effect**

A layer of gases in the atmosphere traps the Sun's energy and keeps Earth warm enough to support life. These are known as greenhouse gases (GHGs). However, human activity has caused levels of these gases to rise, trapping more warmth and so increasing global temperatures.

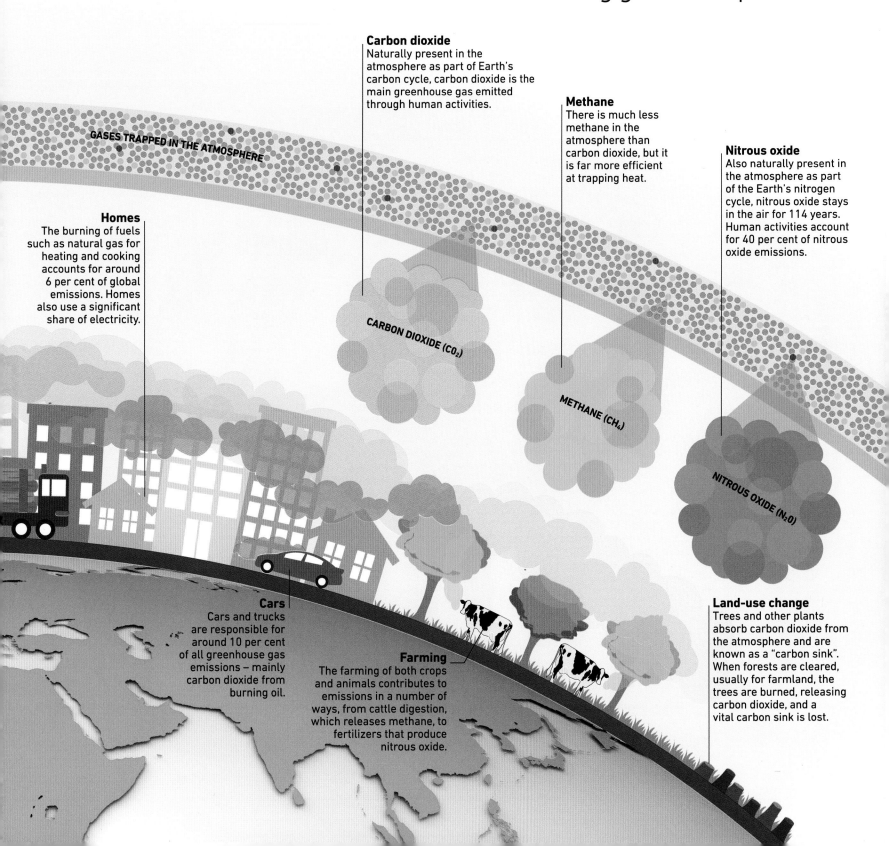

Carbon dioxide
Naturally present in the atmosphere as part of Earth's carbon cycle, carbon dioxide is the main greenhouse gas emitted through human activities.

Methane
There is much less methane in the atmosphere than carbon dioxide, but it is far more efficient at trapping heat.

Nitrous oxide
Also naturally present in the atmosphere as part of the Earth's nitrogen cycle, nitrous oxide stays in the air for 114 years. Human activities account for 40 per cent of nitrous oxide emissions.

GASES TRAPPED IN THE ATMOSPHERE

Homes
The burning of fuels such as natural gas for heating and cooking accounts for around 6 per cent of global emissions. Homes also use a significant share of electricity.

CARBON DIOXIDE (CO_2)

METHANE (CH_4)

NITROUS OXIDE (N_2O)

Cars
Cars and trucks are responsible for around 10 per cent of all greenhouse gas emissions – mainly carbon dioxide from burning oil.

Farming
The farming of both crops and animals contributes to emissions in a number of ways, from cattle digestion, which releases methane, to fertilizers that produce nitrous oxide.

Land-use change
Trees and other plants absorb carbon dioxide from the atmosphere and are known as a "carbon sink". When forests are cleared, usually for farmland, the trees are burned, releasing carbon dioxide, and a vital carbon sink is lost.

The **carbon** cycle

Carbon is an element found in all living things. It flows in cycles through the atmosphere, the oceans, plants, animal life, and rocks. Carbon dioxide (CO_2) is exchanged between the air, the ocean, and the ecosystem through a variety of natural processes that balance each other out. Human activities have upset this balance, causing climate change and ocean acidification.

KEY
Colour-coded arrows show the natural processes and human activities that drive the flow of carbon in the cycle.

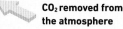 CO_2 **removed from the atmosphere**

CO_2 **added to the atmosphere by natural processes**

CO_2 **added to the atmosphere by human activities**

THE ATMOSPHERE

OCEAN ABSORPTION

PHOTOSYNTHESIS

OCEAN RELEASE

PLANT RESPIRATION

DEFORESTATION

ANIMAL RESPIRATION

Plants
All plants, including trees, absorb CO_2 from the atmosphere, using it to make energy (photosynthesis). Plants also release some CO_2 into the atmosphere by respiration.

Deforestation
Forests act as natural carbon sinks by absorbing and storing more carbon than they emit. Burning trees to clear land not only releases CO_2, it also destroys an important carbon store.

Ocean acidification
Oceans absorb CO_2 from the atmosphere. But if there is too much CO_2, the seawater becomes more acidic. This dissolves the minerals that marine organisms need to build their shells and skeletons.

Warming oceans
As ocean organisms respire, they release CO_2 into the atmosphere. But as seas get warmer, some extra CO_2 is released from the water too.

Animals
CO_2 is released during respiration in all animals – whether they are herbivores, carnivores, or detritivores, which break down dead matter in soil.

Carbon flow

Natural processes such as respiration (the making of energy through food) and burning release CO_2 into the atmosphere, while the oceans and plants absorb it. Because they take in more CO_2 than they release, they are known as "carbon sinks". Human activities have disrupted the carbon cycle, causing more CO_2 to build up in the atmosphere – mostly from burning fossil fuels and deforestation.

Tipping the balance

Growing emissions from human activities have tipped the carbon balance. In total, humans are responsible for adding 10 billion tonnes of CO_2 to the atmosphere every year.

Human sources of CO_2

Natural carbon sinks

Natural sources of CO_2

AIR TRAVEL

INDUSTRIAL EMISSIONS

HOUSEHOLD EMISSIONS

VOLCANIC ACTIVITY

TRANSPORT

FARMING

Flight
Planes are powered by burning kerosene, produced from oil.

Volcanoes
Erupting volcanoes, on land and under the sea, generate CO_2, but the amount they produce is just one per cent of the emissions released by human activities.

Industry
Fossil fuels such as coal power many industries.

Homes
Coal, natural gas, and oil are burned to generate electricity for homes. Fossil fuels are also used for cooking and heating.

Farming
As cattle and other farm animals digest their food they produce methane, a gas that contains carbon. Growing rice also releases methane into the atmosphere.

Road transport
Cars and trucks are powered by burning fossil fuels such as petrol and diesel, produced from oil.

Fossil fuels
Over millions of years, buried plant and animal remains are transformed by pressure and heat into coal, oil, and gas. These are extracted and used as fuels. When they are burned, they release CO_2.

Dead matter
When animals and plants produce waste or die, they add dead matter (containing carbon) to the soil.

Microbes
Tiny microbes in the soil produce CO_2 during respiration as they break down dead matter.

What's your carbon footprint?

Carbon footprints measure the greenhouse gases (GHGs) released by an activity, the making and supply of a product, or you! A person's carbon footprint is based on emissions from their lifestyle.

30 tonnes GHGs

QATAR
One of the wealthiest nations in the world, Qatar also has one of the largest carbon footprints. A big part of this comes from the fossil fuels the are used to power the process of removing salt from sea water to drink.

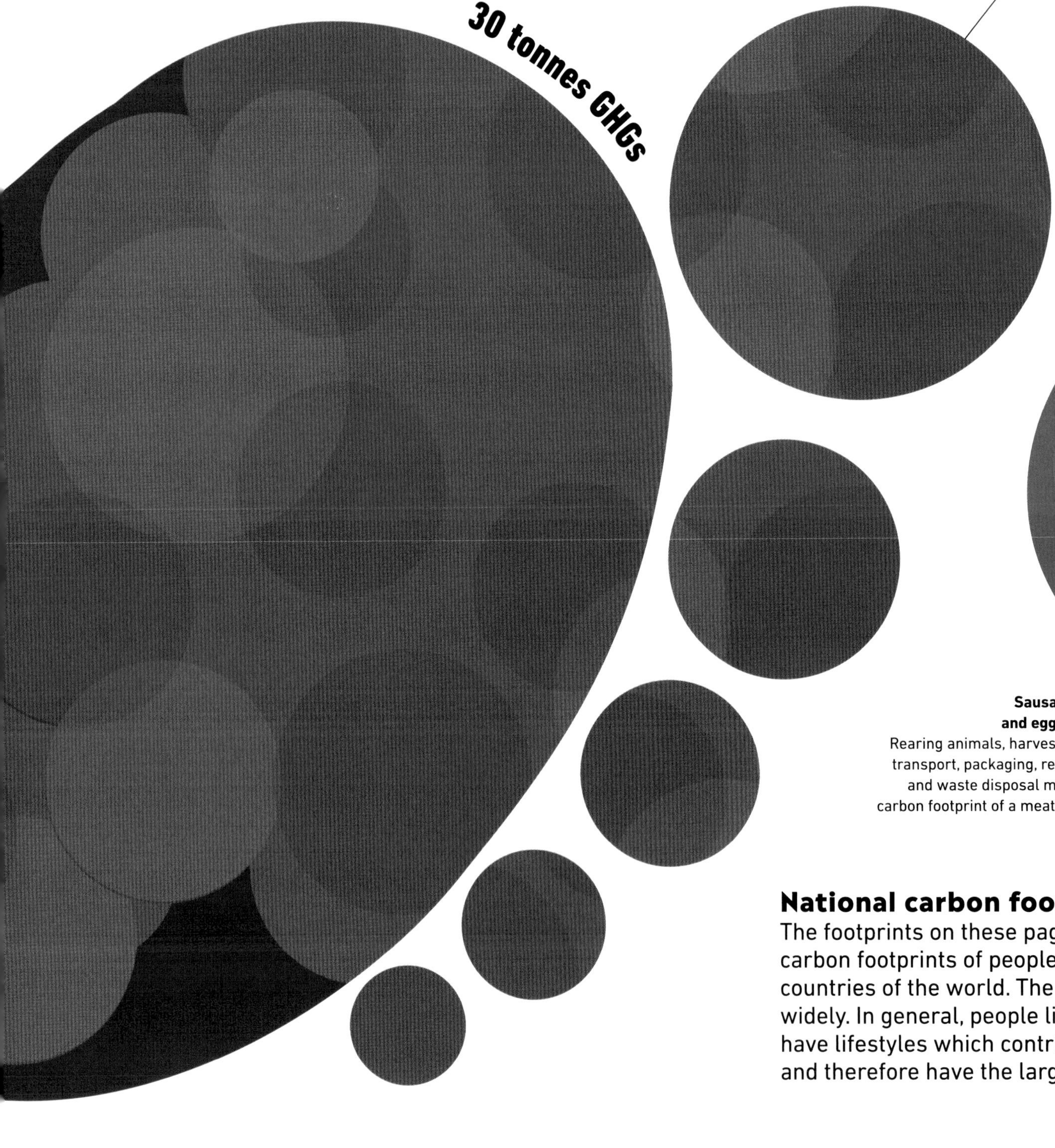

Watching TV
Using electricity to watch TV is a relatively small part of an individual's footprint, but this is an activity that is more popular in wealthier countries.

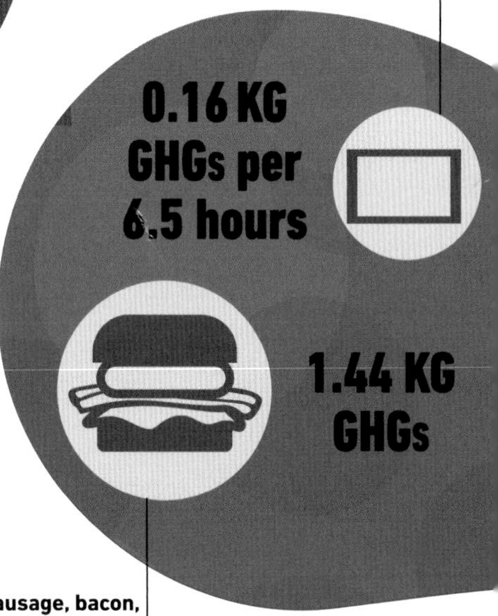

0.16 KG GHGs per 6.5 hours

1.44 KG GHGs

Sausage, bacon, and egg sandwich
Rearing animals, harvesting crops, transport, packaging, refrigeration and waste disposal make up the carbon footprint of a meat sandwich.

National carbon footprint
The footprints on these pages show average annual carbon footprints of people in a range of different countries of the world. The size of the footprints varies widely. In general, people living in wealthier nations have lifestyles which contribute the most emissions, and therefore have the largest carbon footprints.

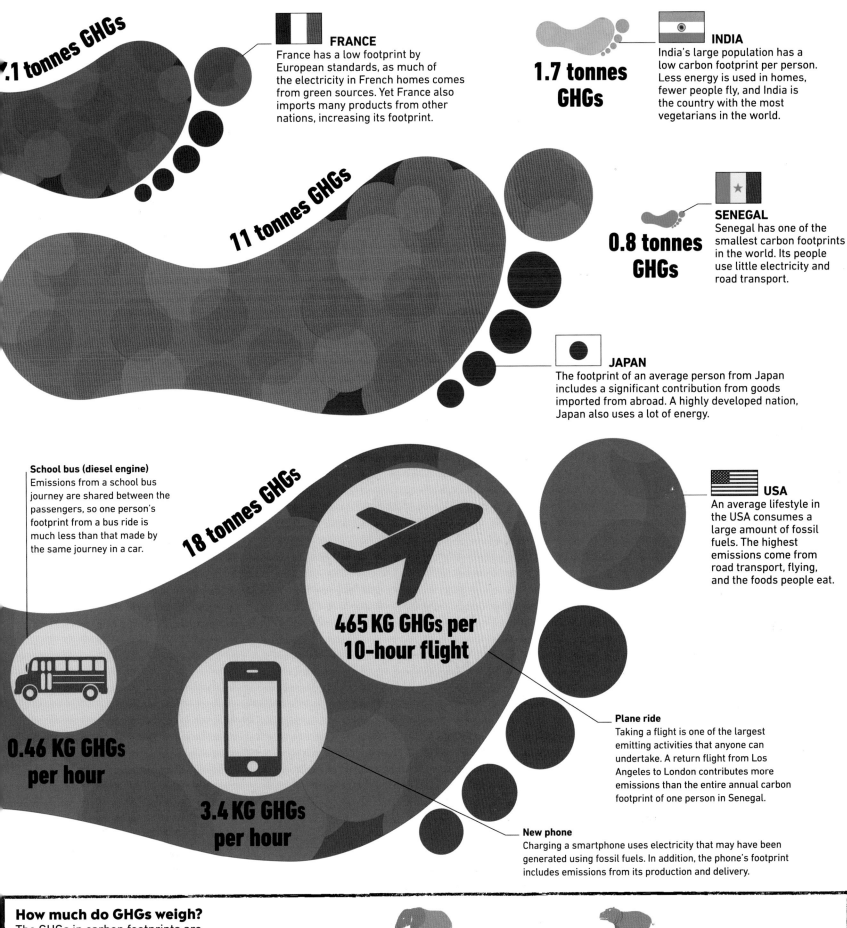

.1 tonnes GHGs

FRANCE
France has a low footprint by European standards, as much of the electricity in French homes comes from green sources. Yet France also imports many products from other nations, increasing its footprint.

INDIA
India's large population has a low carbon footprint per person. Less energy is used in homes, fewer people fly, and India is the country with the most vegetarians in the world.

1.7 tonnes GHGs

11 tonnes GHGs

SENEGAL
Senegal has one of the smallest carbon footprints in the world. Its people use little electricity and road transport.

0.8 tonnes GHGs

JAPAN
The footprint of an average person from Japan includes a significant contribution from goods imported from abroad. A highly developed nation, Japan also uses a lot of energy.

School bus (diesel engine)
Emissions from a school bus journey are shared between the passengers, so one person's footprint from a bus ride is much less than that made by the same journey in a car.

18 tonnes GHGs

465 KG GHGs per 10-hour flight

USA
An average lifestyle in the USA consumes a large amount of fossil fuels. The highest emissions come from road transport, flying, and the foods people eat.

0.46 KG GHGs per hour

3.4 KG GHGs per hour

Plane ride
Taking a flight is one of the largest emitting activities that anyone can undertake. A return flight from Los Angeles to London contributes more emissions than the entire annual carbon footprint of one person in Senegal.

New phone
Charging a smartphone uses electricity that may have been generated using fossil fuels. In addition, the phone's footprint includes emissions from its production and delivery.

How much do GHGs weigh?
The GHGs in carbon footprints are given as weight – but what do those numbers mean? One way to understand the weight is to visualize the figures as familiar objects. The weight of an average person's carbon footprint in France, for instance, is equal to the weight of three hippos.

USA   FRANCE   INDIA

Studying climate change

Using detailed observations of conditions today, climate scientists study how the climate is changing as greenhouse gases (GHGs) build up in the atmosphere. By gathering this data, scientists can also predict how the climate will change in the future if concentrations of GHGs increase.

Gathering data

Climate scientists collect and analyse data from a huge range of sources. They monitor weather conditions at Earth's surface and in the upper atmosphere, observe ocean temperature and currents, and track surface features such as ice cover. Networks of sensors transmit data from single locations, while measurements taken remotely by satellites can help to fill in any blanks. Taken together, this information shows how Earth's climate has changed over time.

Eye in the sky
Satellites orbiting Earth measure many conditions remotely, such as air temperature, cloud cover, and levels of pollution. Observation satellites scan the surface, allowing scientists to track how deserts are growing and areas of sea ice cover are receding.

Weather ship
More than 70 per cent of Earth is covered by water, so to get a complete picture of the weather at the surface of the ocean, ships are used, as well as buoys and platforms. They collect information about air pressure, wind speed and direction, temperature, and humidity.

Monitoring the oceans
A network of thousands of drifting buoys and stationary platforms measure weather conditions and relay information to weather stations on land or at sea. Buoys also measure ocean currents and wave height, while probes called Argo floats sink and rise in the water itself to monitor ocean temperature and salt levels at different depths. All this contributes to our understanding of how climate change is affecting the oceans.

Radar wind profilers
On land, radar wind profilers use radio waves to measure wind speed and direction.

Weather radar
Radar towers measure rain by sending out bursts of radio waves. A pulse comes back when the waves hit droplets of water in the air.

Predicting future climate change

Studying the climate shows how GHG emissions must be reduced to limit temperature rise. The labels on this graph show the predicted temperature increase for a number of emissions scenarios, including how emissions need to be reduced to keep warming below the 2°C (3.6°F) target set in the Paris Agreement.

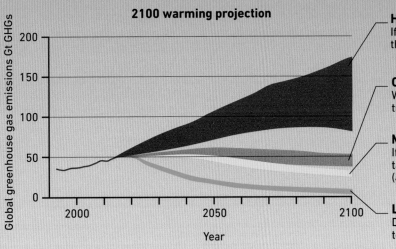

2100 warming projection

Global greenhouse gas emissions Gt GHGs

Year

High emissions
If no action is taken, warming of more than 4°C (7.2°F) is likely by 2100.

Current path
With current policies, the world is on track to warm by up to 3.2°C (5.8°F).

Meeting targets
If every country meets its climate targets, warming will be 2.5–2.8°C (4.5–5°F).

Low emissions
Drastic action would be needed to limit warming to 2°C (3.6°F).

Radiosonde
Rising high into the upper atmosphere, a weather balloon carries an instrument called a radiosonde. This miniature weather station measures pressure, wind, temperature, and humidity at different altitudes, and sends the data back to a base station on the ground.

Weather satellite
Satellites in geostationary orbit remain over the same location on Earth's surface but orbit high above the planet so they can see the whole globe. They monitor hurricanes and other extreme weather events, which are becoming more frequent as temperatures on Earth increase.

Dropsonde
Similar to a radiosonde, a dropsonde measures conditions in the atmosphere. Dropped from an aeroplane and drifting down to the surface by parachute, they are used over areas such as the open ocean where it is not possible to launch a balloon.

Aircraft
Many commercial airliners carry equipment that measures weather conditions and sends data to weather stations on the ground. Dedicated meteorological aeroplanes are used to study cloud systems and aerosols (liquid and solid particles suspended in air) in the atmosphere.

Base station
Weather stations on the ground measure air temperature and pressure, wind speed and direction, precipitation, and humidity. Scientists use the data from all the different sources to build climate models – computer programs that simulate Earth's complex climate. They use these models to find out how the climate will respond to changing levels of greenhouse gases in the atmosphere.

The causes of climate change

Why is the climate changing?

Human activities – from farming and manufacturing to transport and heating – emit greenhouse gases (GHGs). Over the past 250 years, the scale and intensity of these activities have dramatically accelerated and so, as a result, has climate change.

Power

Most of the energy used to generate electricity is still produced from fossil fuels, such as coal and gas.

Growing demand

As populations grow larger and wealthier, countries consume more and more food and energy.

Transport

Cars and aeroplanes use oil products, such as petrol and diesel, to power internal combustion engines.

Fossil fuel burning

To provide power for homes, businesses, and industries, coal, oil, and natural gas are burned in power plants. This generates carbon dioxide (CO_2), a greenhouse gas.

Industry

Manufacturing goods, such as clothing or toys, generates emissions across all stages of the production process.

Foods

Growing food and cattle farming emit GHGs, including harmful nitrous oxide and methane.

Deforestation

Forests remove CO_2 from the atmosphere, but large areas have been cleared to make room for agriculture.

Carbon dioxide (CO₂)

Human activity is pumping the heat-trapping CO_2 that causes climate change into the atmosphere. This chart shows the steady increase in CO_2 concentration in Earth's atmosphere over the last 60 years.

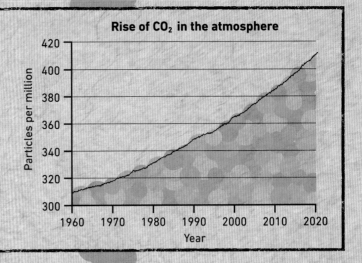

Rise of CO₂ in the atmosphere

Human carbon footprint

The combination of all these activities has caused the dramatic build-up of harmful GHGs in the atmosphere, warming our planet. This impact can be described as the human carbon footprint.

Population growth

There are eight times as many people on Earth today as there were two centuries ago. As countries industrialize and people consume more goods and services, their greenhouse gas emissions increase.

Growth over time

This sequence of world maps shows how the global population has grown over time and how this is speeding up. For thousands of years, the population grew very slowly. Things started to take off during the Industrial Revolution, around the same time that people began to burn the fossil fuels that emit greenhouse gases.

The first farmers
When humans started farming the land, the world population was less than that of many large cities today.

10,000 BCE
4,000,000

Slow growth
The population grew slowly, taking another 11,500 years to reach just under half a billion people.

1500
480,000,000

A farmer's life
Before industrialization, most people were farmers. Many died young, often under the age of 40.

1750
770,000,000

Where in the world

As this map shows, people are not evenly distributed around the world – population density varies hugely across countries. CO_2 emissions vary massively across populations too, depending largely on economic prosperity.

Europe
With a population of 700 million, this densely populated and highly developed region has historically been the biggest emitter of CO_2 but now emits less than the USA and China.

USA
Home to 327 million people – 5 per cent of the world's population – the wealthy USA emits 15 per cent of global CO_2, making it the biggest emitter after China.

Nigeria
Africa's most populous country, Nigeria is home to 196 million people. Like other African countries, it has low incomes and low levels of CO_2 emissions: 16 per cent of the world's population lives on the continent, but it accounts for just 4 per cent of global CO_2.

How many Earths?
Humans consume so much that we are using more than Earth can produce. In fact, according to one estimate, we need the resources of 1.75 Earths to sustain current consumption. However, some countries with large populations consume relatively few resources. In fact, if everyone lived like the average citizen of India – the world's second most populous country – on a median income of just $616, we would only need half of the world's resources.

| Qatar | Luxembourg | USA | Australia | Germany | China | Brazil | Ecuador | India | Burundi |

THE POPULATION OF EARTH IS PREDICTED TO STABILIZE AT 11 BILLION BY 2100

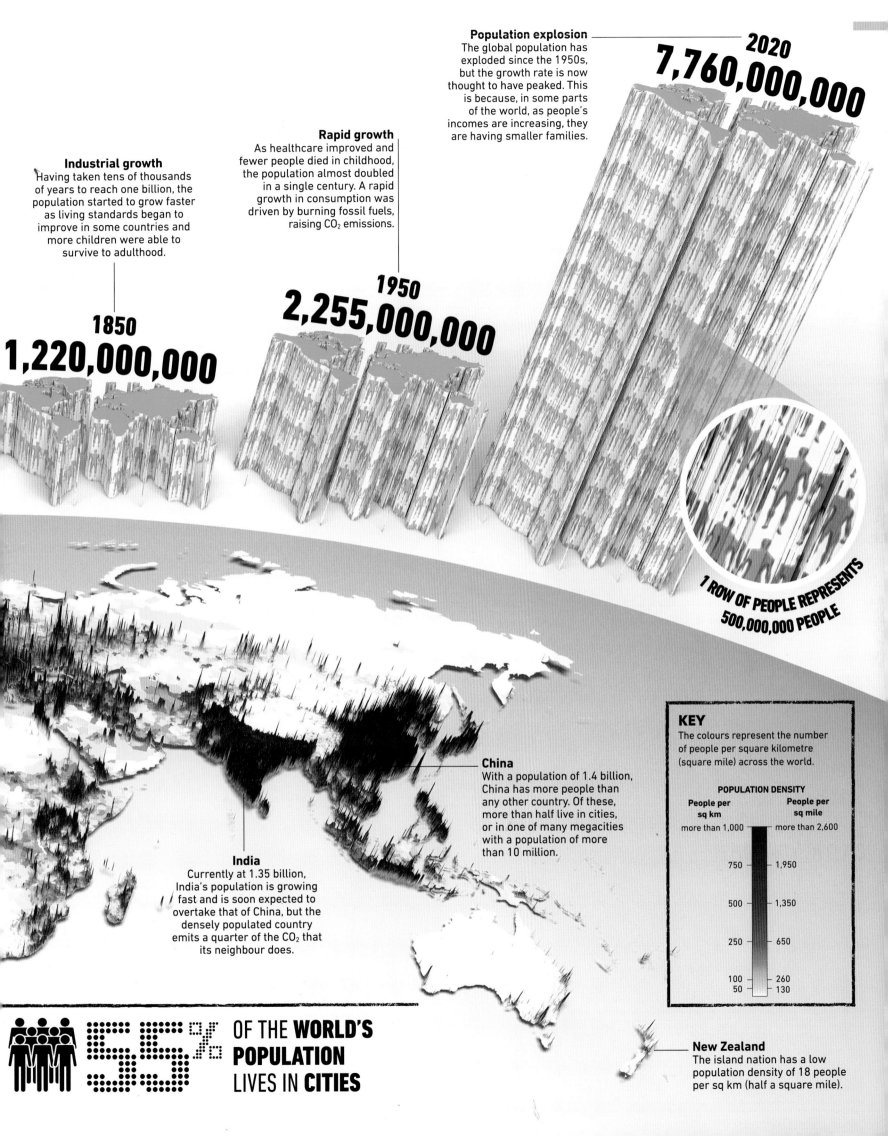

Population explosion
The global population has exploded since the 1950s, but the growth rate is now thought to have peaked. This is because, in some parts of the world, as people's incomes are increasing, they are having smaller families.

2020
7,760,000,000

Rapid growth
As healthcare improved and fewer people died in childhood, the population almost doubled in a single century. A rapid growth in consumption was driven by burning fossil fuels, raising CO_2 emissions.

1950
2,255,000,000

Industrial growth
Having taken tens of thousands of years to reach one billion, the population started to grow faster as living standards began to improve in some countries and more children were able to survive to adulthood.

1850
1,220,000,000

1 ROW OF PEOPLE REPRESENTS 500,000,000 PEOPLE

China
With a population of 1.4 billion, China has more people than any other country. Of these, more than half live in cities, or in one of many megacities with a population of more than 10 million.

India
Currently at 1.35 billion, India's population is growing fast and is soon expected to overtake that of China, but the densely populated country emits a quarter of the CO_2 that its neighbour does.

KEY
The colours represent the number of people per square kilometre (square mile) across the world.

POPULATION DENSITY

People per sq km	People per sq mile
more than 1,000	more than 2,600
750	1,950
500	1,350
250	650
100	260
50	130

55% OF THE **WORLD'S POPULATION** LIVES IN **CITIES**

New Zealand
The island nation has a low population density of 18 people per sq km (half a square mile).

Burning fossil fuels

Using fossil fuels – coal, oil, and gas – to generate power creates the carbon dioxide (CO_2) emissions that cause climate change. The biggest emitter today is China, but over the past century most greenhouse gases have come from the developed economies of Europe and North America.

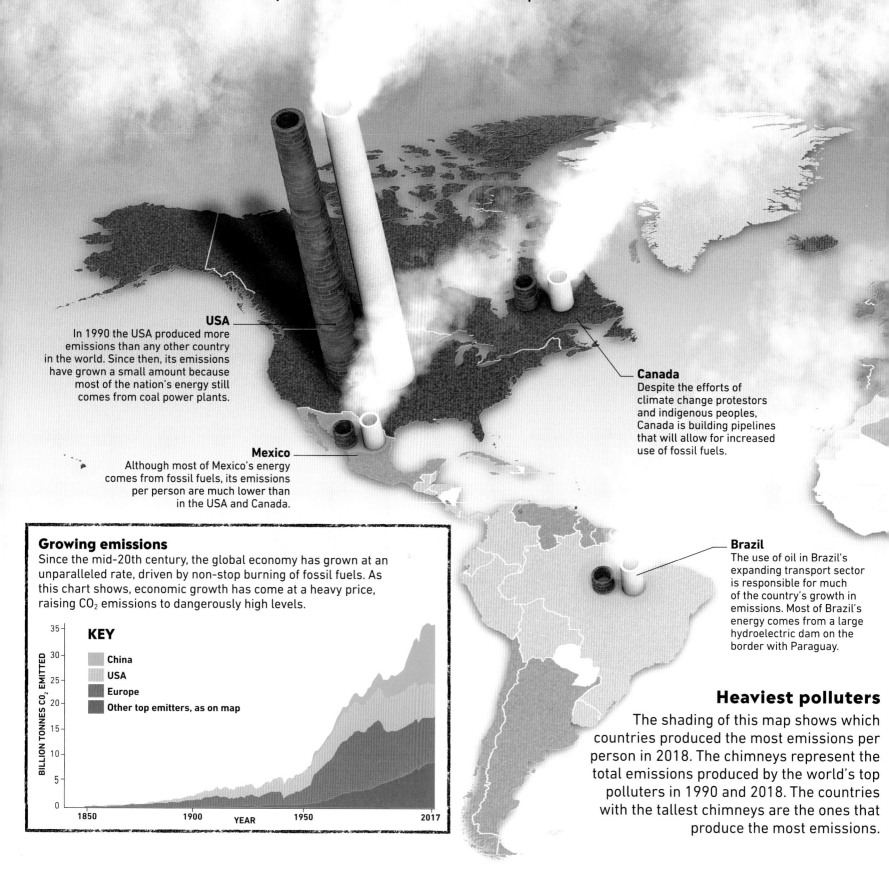

USA
In 1990 the USA produced more emissions than any other country in the world. Since then, its emissions have grown a small amount because most of the nation's energy still comes from coal power plants.

Canada
Despite the efforts of climate change protestors and indigenous peoples, Canada is building pipelines that will allow for increased use of fossil fuels.

Mexico
Although most of Mexico's energy comes from fossil fuels, its emissions per person are much lower than in the USA and Canada.

Brazil
The use of oil in Brazil's expanding transport sector is responsible for much of the country's growth in emissions. Most of Brazil's energy comes from a large hydroelectric dam on the border with Paraguay.

Growing emissions
Since the mid-20th century, the global economy has grown at an unparalleled rate, driven by non-stop burning of fossil fuels. As this chart shows, economic growth has come at a heavy price, raising CO_2 emissions to dangerously high levels.

KEY
- China
- USA
- Europe
- Other top emitters, as on map

Y-axis: BILLION TONNES CO_2 EMITTED — 0, 5, 10, 15, 20, 25, 30, 35

X-axis: YEAR — 1850, 1900, 1950, 2017

Heaviest polluters

The shading of this map shows which countries produced the most emissions per person in 2018. The chimneys represent the total emissions produced by the world's top polluters in 1990 and 2018. The countries with the tallest chimneys are the ones that produce the most emissions.

20% OF GREENHOUSE **GASES COME FROM INDUSTRY**

CARBON DIOXIDE MAKES UP **74% OF ALL GREENHOUSE GAS** EMISSIONS

KEY

The map shows CO_2 emissions per capita of all countries in the world.

- below 1.0t (tonnes per year)
- 1.0–2.5 t
- 2.5–5.0 t
- 5.0–10.0 t
- 10.0–15.0 t
- above 15.0 t
- no data

The chimneys show the growth in total emissions.

1990   2018

China
With the biggest population in the world and a rapidly growing economy, China's use of fossil fuels has vastly expanded and the country is now responsible for 26 per cent of the world's emissions. However, its emissions per person are relatively moderate.

Iran
Iran has the world's largest natural gas and its second-largest oil reserves and, like many resource-rich countries, its emissions have grown over the past 30 years.

Russia
From being one of the biggest emitters in the last century, Russia has actually decreased its fossil fuel use over the last 30 years.

Germany
Although it is still Europe's biggest emitter, Germany has invested in renewable energy and its emissions have decreased since 1990.

South Korea
The word's 10th-highest emitter of greenhouse gases, South Korea's economy has grown rapidly since 1990, rising hand-in-hand with the country's CO_2 emissions.

Japan
For many years, Japan has been a major importer of coal, oil, and natural gas. Although the country's plans to cut fossil fuels were set back by the 2011 Fukushima nuclear disaster, its emissions have been decreasing since 2013.

Saudi Arabia
With vast oil reserves, Saudi Arabia uses the most fossil fuels of any country in the Middle East.

India
India has the world's second largest population, but even though its industry has grown rapidly in the last 30 years, it still produces comparatively few emissions per person.

South Africa
Of the coal consumed in Africa, 92 per cent comes from South African mines. South Africa itself burns a lot of coal and says it is unlikely to shift to a different source of energy.

Australia
Australia uses coal to produce 85 per cent of its energy. Although it is a large country, it has a relatively small population, so its emissions per person are among the world's highest.

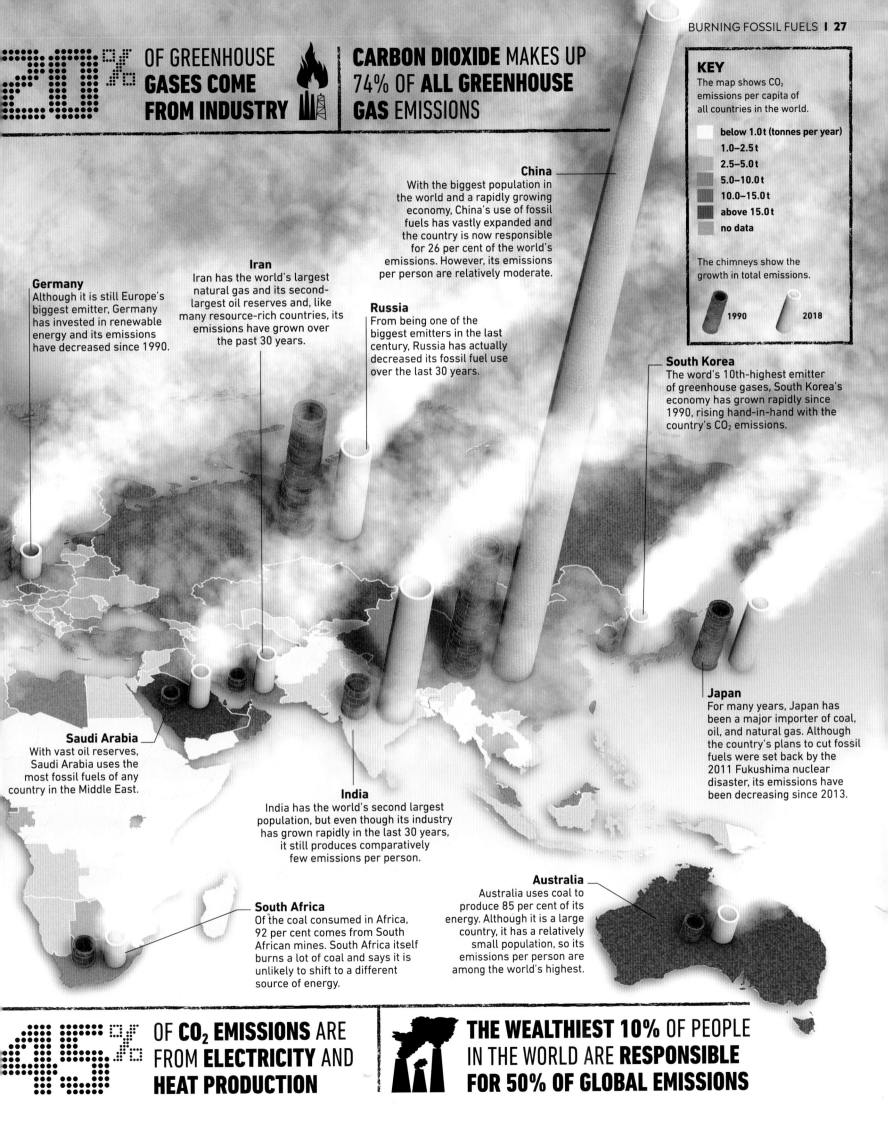

25% OF **CO_2 EMISSIONS** ARE FROM **ELECTRICITY** AND **HEAT PRODUCTION**

THE WEALTHIEST 10% OF PEOPLE IN THE WORLD ARE **RESPONSIBLE FOR 50% OF GLOBAL EMISSIONS**

In November 2019 the air quality in New Delhi was so poor that the government declared a public health emergency.

Air pollution

This image shows the India Gate monument in New Delhi, India, before and during the COVID-19 lockdown. In November 2019, thick smog surrounded the monument, as the air was choked with toxic emissions from motor vehicles, industry, and agriculture. One of the world's most polluted cities, New Delhi's Air Quality Index (AQI) – a measure of harmful particles in the air, where a reading between 0 and 50 is considered good – regularly measures above 200 and has been known to exceed 900. When the country went into lockdown in late March 2020, during the COVID-19 pandemic, industries shut down and cars disappeared from the roads. Air quality improved immediately. The effects of the lockdown demonstrated the rewards that can be gained from switching to less polluting fuels.

20 APRIL 2020

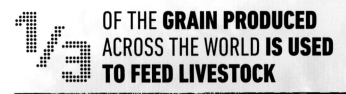

1/3 OF THE **GRAIN PRODUCED** ACROSS THE WORLD **IS USED TO FEED LIVESTOCK**

70 BILLION ANIMALS ARE RAISED EVERY YEAR **FOR** HUMAN **CONSUMPTION**

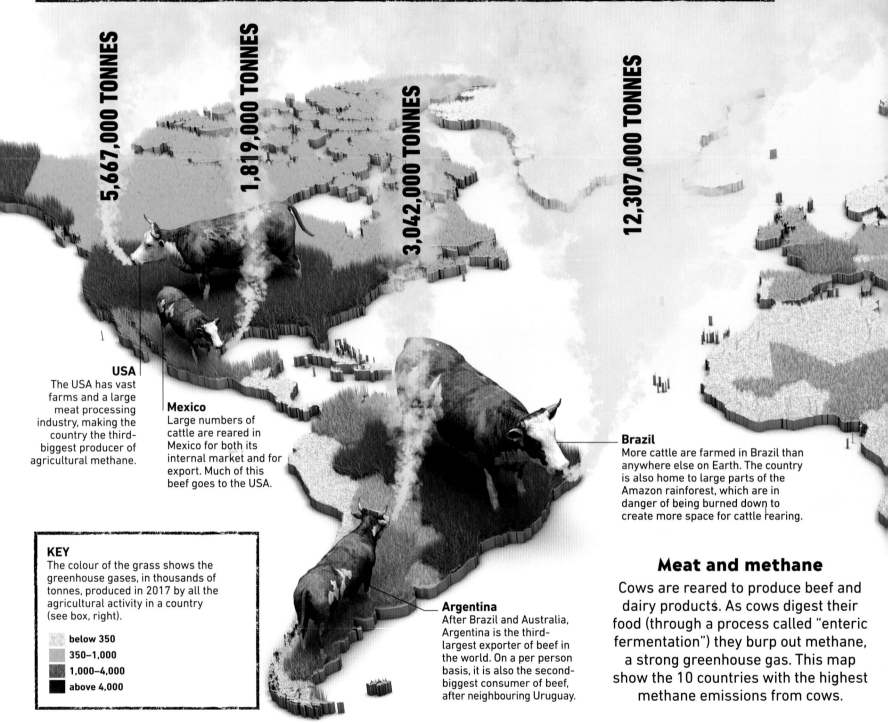

5,667,000 TONNES

1,819,000 TONNES

3,042,000 TONNES

12,307,000 TONNES

USA
The USA has vast farms and a large meat processing industry, making the country the third-biggest producer of agricultural methane.

Mexico
Large numbers of cattle are reared in Mexico for both its internal market and for export. Much of this beef goes to the USA.

Brazil
More cattle are farmed in Brazil than anywhere else on Earth. The country is also home to large parts of the Amazon rainforest, which are in danger of being burned down to create more space for cattle rearing.

KEY
The colour of the grass shows the greenhouse gases, in thousands of tonnes, produced in 2017 by all the agricultural activity in a country (see box, right).

- below 350
- 350–1,000
- 1,000–4,000
- above 4,000

Argentina
After Brazil and Australia, Argentina is the third-largest exporter of beef in the world. On a per person basis, it is also the second-biggest consumer of beef, after neighbouring Uruguay.

Meat and methane
Cows are reared to produce beef and dairy products. As cows digest their food (through a process called "enteric fermentation") they burp out methane, a strong greenhouse gas. This map show the 10 countries with the highest methane emissions from cows.

Farming emissions

Agriculture plays a huge part in the build-up of global greenhouse gases. Its main impact is from the cutting down of forests, which take carbon dioxide (CO_2) from the atmosphere, to clear space for farmland. Another major factor is meat farming, which uses land both for grazing and producing animal feed.

1,376,000 TONNES

6,576,000 TONNES

1,604,000 TONNES

4,170,000 TONNES

1,616,000 TONNES

2.06 ... 0 TONNES

Russia
With large amounts of land turned over to cattle rearing, Russia is capable of producing beef and milk for export to countries as far away as Morocco in North Africa.

China
As well as importing beef from around the world, China has large ranches of its own. As the Chinese economy has grown, its citizens have started consuming more beef and milk.

Pakistan
The fourth-largest producer of milk in the world, Pakistan is reported to have more than 24.2 million cows. It also produces beef for export to China.

India
The largest producer of dairy products in the world, India is home to more than 350 million cows and water buffalo, which together produce 19 percent of the planet's milk.

Ethiopia
The largest producer of beef in Africa, Ethiopia processes about 70,000 cattle a year for consumption locally and export.

Australia
There are about 25 million cows in Australia, and cattle are reared in every state in the country. Of all farms in the country, 55 per cent are beef farms.

Sources of agricultural emissions

Farming is responsible for three main greenhouse gases. In addition to the methane produced by enteric fermentation and rice, fertilizers and intensive cultivation of soil release nitrous oxide, a gas that is very effective at warming the atmosphere. Agriculture is also responsible for large amounts of CO_2. As forests are cleared to create more space for crops and animals, trees are burned, releasing CO_2. Fewer trees also mean less CO_2 is absorbed from the atmosphere.

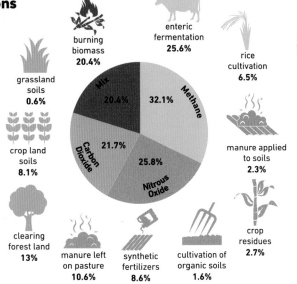

burning biomass
20.4%

enteric fermentation
25.6%

rice cultivation
6.5%

grassland soils
0.6%

Mix
20.4%

Methane
32.1%

crop land soils
8.1%

Carbon Dioxide
21.7%

Nitrous Oxide
25.8%

manure applied to soils
2.3%

clearing forest land
13%

manure left on pasture
10.6%

synthetic fertilizers
8.6%

cultivation of organic soils
1.6%

crop residues
2.7%

AN **AREA** THE SIZE OF **SWEDEN** HAS BEEN **CLEARED** FOR CATTLE PASTURE

IN THE AMAZON RAINFOREST

SINCE THE **1960S**

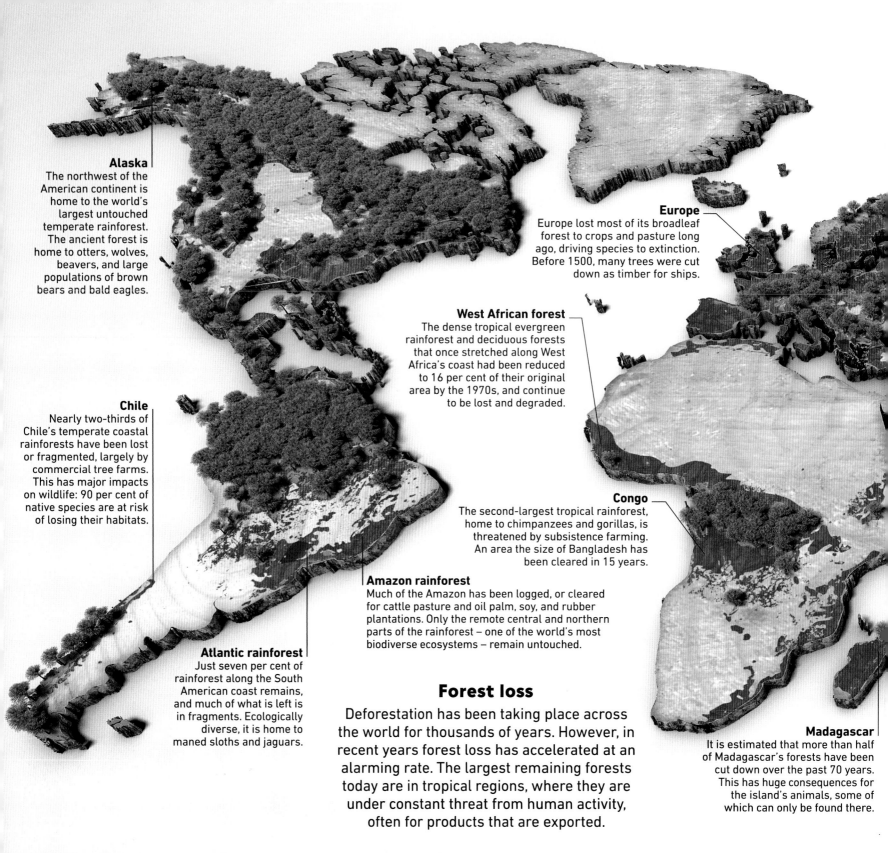

Alaska
The northwest of the American continent is home to the world's largest untouched temperate rainforest. The ancient forest is home to otters, wolves, beavers, and large populations of brown bears and bald eagles.

Europe
Europe lost most of its broadleaf forest to crops and pasture long ago, driving species to extinction. Before 1500, many trees were cut down as timber for ships.

West African forest
The dense tropical evergreen rainforest and deciduous forests that once stretched along West Africa's coast had been reduced to 16 per cent of their original area by the 1970s, and continue to be lost and degraded.

Chile
Nearly two-thirds of Chile's temperate coastal rainforests have been lost or fragmented, largely by commercial tree farms. This has major impacts on wildlife: 90 per cent of native species are at risk of losing their habitats.

Congo
The second-largest tropical rainforest, home to chimpanzees and gorillas, is threatened by subsistence farming. An area the size of Bangladesh has been cleared in 15 years.

Amazon rainforest
Much of the Amazon has been logged, or cleared for cattle pasture and oil palm, soy, and rubber plantations. Only the remote central and northern parts of the rainforest – one of the world's most biodiverse ecosystems – remain untouched.

Atlantic rainforest
Just seven per cent of rainforest along the South American coast remains, and much of what is left is in fragments. Ecologically diverse, it is home to maned sloths and jaguars.

Forest loss

Deforestation has been taking place across the world for thousands of years. However, in recent years forest loss has accelerated at an alarming rate. The largest remaining forests today are in tropical regions, where they are under constant threat from human activity, often for products that are exported.

Madagascar
It is estimated that more than half of Madagascar's forests have been cut down over the past 70 years. This has huge consequences for the island's animals, some of which can only be found there.

Disappearing forests

Forests are vital to Earth's climate. They are known as "carbon sinks" because of trees' ability to capture carbon dioxide gas from the air. But they are disappearing fast – by 2011, half of the world's original forests had been cut down by humans, mostly to create space for new farmland.

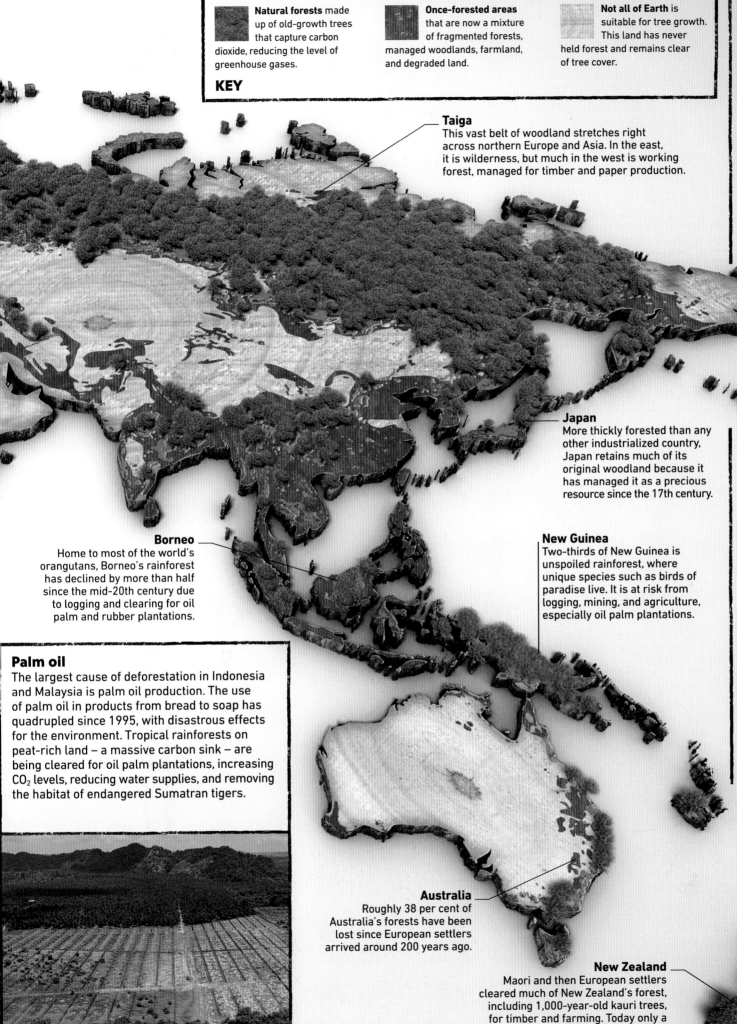

KEY

Natural forests made up of old-growth trees that capture carbon dioxide, reducing the level of greenhouse gases.

Once-forested areas that are now a mixture of fragmented forests, managed woodlands, farmland, and degraded land.

Not all of Earth is suitable for tree growth. This land has never held forest and remains clear of tree cover.

37%
OF THE WORLD'S HABITABLE LAND IS FOREST

Taiga
This vast belt of woodland stretches right across northern Europe and Asia. In the east, it is wilderness, but much in the west is working forest, managed for timber and paper production.

Japan
More thickly forested than any other industrialized country, Japan retains much of its original woodland because it has managed it as a precious resource since the 17th century.

BETWEEN 1990 AND 2015 WE LOST

880
FOOTBALL PITCHES WORTH OF FOREST EVERY HOUR

Borneo
Home to most of the world's orangutans, Borneo's rainforest has declined by more than half since the mid-20th century due to logging and clearing for oil palm and rubber plantations.

New Guinea
Two-thirds of New Guinea is unspoiled rainforest, where unique species such as birds of paradise live. It is at risk from logging, mining, and agriculture, especially oil palm plantations.

Palm oil
The largest cause of deforestation in Indonesia and Malaysia is palm oil production. The use of palm oil in products from bread to soap has quadrupled since 1995, with disastrous effects for the environment. Tropical rainforests on peat-rich land – a massive carbon sink – are being cleared for oil palm plantations, increasing CO_2 levels, reducing water supplies, and removing the habitat of endangered Sumatran tigers.

Australia
Roughly 38 per cent of Australia's forests have been lost since European settlers arrived around 200 years ago.

New Zealand
Maori and then European settlers cleared much of New Zealand's forest, including 1,000-year-old kauri trees, for timber and farming. Today only a quarter of the country is native forest.

Between 1990 and 2015 Indonesia lost around a quarter of its total forest cover.

Forest devastation

In Papua province, Indonesia, tropical forest is cleared to make way for oil palm saplings. The fruit of the oil palm is pressed to make palm oil, the world's most consumed oil – more than 70 million tonnes of it are produced every year, with Indonesia the largest supplier. Huge swathes of the country's native forests are cleared each year to make way for vast plantations, releasing millions of tonnes of carbon into the atmosphere in the process. Some of the forests grow on swampy peatland, which itself is incredibly rich in carbon. When the swamps are drained to grow oil palms, greenhouse gases locked in the peat are released into the atmosphere. In spite of a government ban, deforestation in Indonesia continues.

Road transport

Cars and trucks rely on burning the petrol and diesel that comes from oil, a fossil fuel. Together, road vehicles account for more than 10 per cent of global carbon dioxide (CO_2) emissions. After decades behind the wheel, the world's driving habit is hard to break.

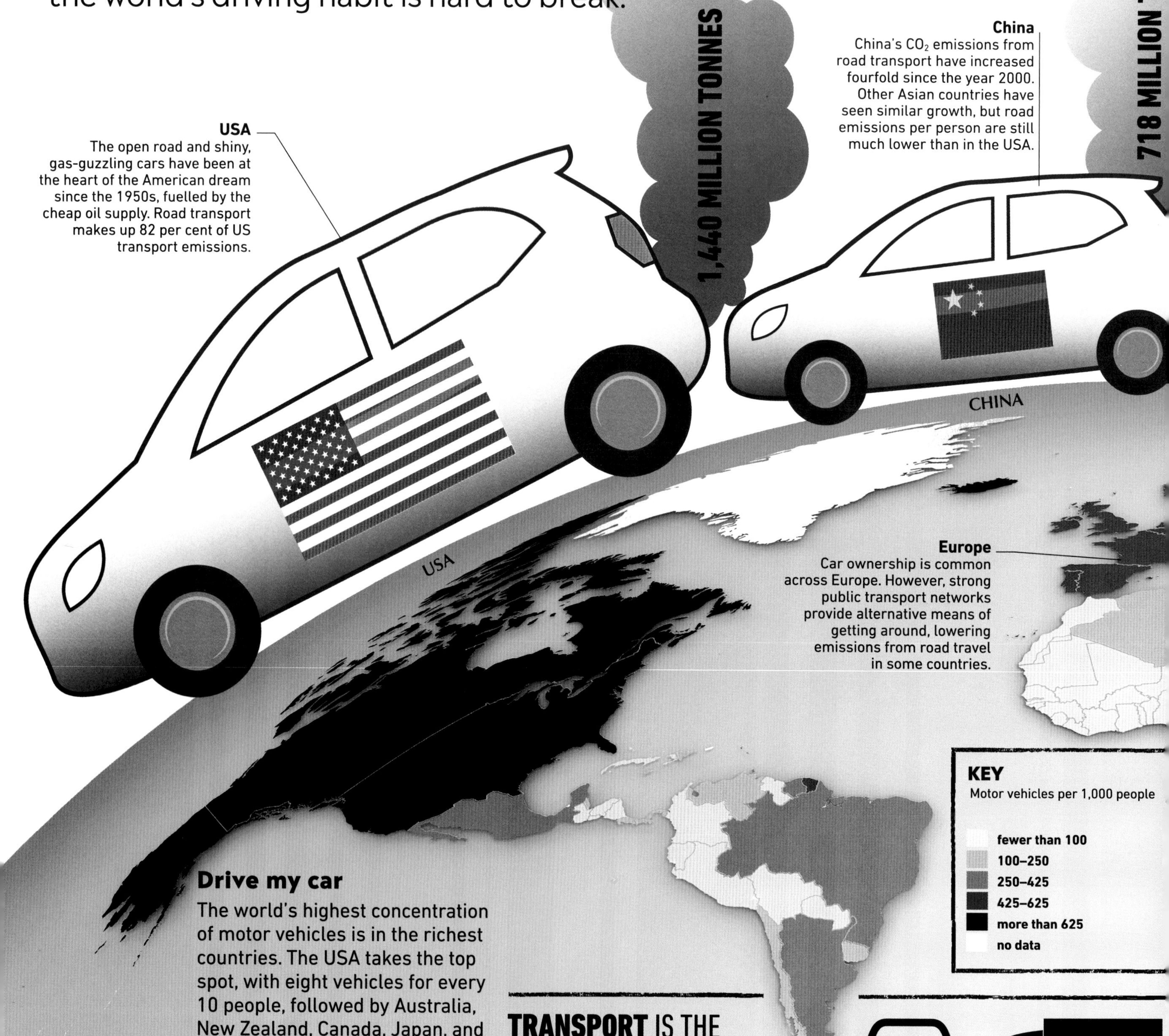

1,440 MILLION TONNES

718 MILLION TONNES

China
China's CO_2 emissions from road transport have increased fourfold since the year 2000. Other Asian countries have seen similar growth, but road emissions per person are still much lower than in the USA.

USA
The open road and shiny, gas-guzzling cars have been at the heart of the American dream since the 1950s, fuelled by the cheap oil supply. Road transport makes up 82 per cent of US transport emissions.

CHINA

USA

Europe
Car ownership is common across Europe. However, strong public transport networks provide alternative means of getting around, lowering emissions from road travel in some countries.

KEY
Motor vehicles per 1,000 people

	fewer than 100
	100–250
	250–425
	425–625
	more than 625
	no data

Drive my car
The world's highest concentration of motor vehicles is in the richest countries. The USA takes the top spot, with eight vehicles for every 10 people, followed by Australia, New Zealand, Canada, Japan, and much of Europe. In China, which has more than 300 million cars, there are two for every 10 people.

TRANSPORT IS THE FASTEST-GROWING SOURCE OF GLOBAL EMISSIONS

Car emissions

Globally, carbon dioxide (CO_2) emissions from road transport account for nearly three-quarters of all CO_2 emitted by transport – but some countries belch out far more from their vehicle exhaust pipes than others. Passenger cars are responsible for 60 per cent of greenhouse gases on the road, freight lorries and vans for most of the rest.

How should I travel?

Using motorized transport, whether for a holiday, to visit family, or for work, increases our carbon footprint by different amounts, and it is possible to compare how much carbon dioxide (CO_2) is emitted if we go by plane, car, or train. For example, one person flying on Europe's busiest route – from Paris to Toulouse – is responsible for 28 times as much CO_2 as a person taking the same trip by train. A solo driver on the same route is responsible for three times as much CO_2 as a person sharing the drive with three passengers.

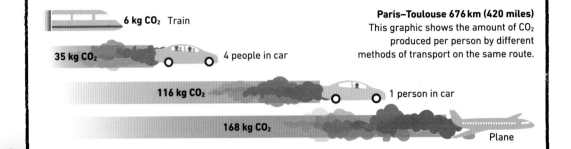

6 kg CO_2 Train

35 kg CO_2 4 people in car

116 kg CO_2 1 person in car

168 kg CO_2 Plane

Paris–Toulouse 676 km (420 miles)
This graphic shows the amount of CO_2 produced per person by different methods of transport on the same route.

265 MILLION TONNES

185 MILLION TONNES

185 MILLION TONNES

158 MILLION TONNES

149 MILLION TONNES

147 MILLION TONNES

136.7 MILLION TONNES

131 MILLION TONNES

INDIA

BRAZIL

JAPAN

GERMANY

RUSSIA

MEXICO

CANADA

IRAN

Germany
It is perhaps no surprise that Germany, renowned for car brands such as Porsche and Volkswagen, is a nation of drivers: 95 per cent of its transport emissions come from road traffic. One-third of these emissions are caused by freight lorries.

Iran
Iran has rich natural oil reserves, which it uses to fuel its road transport – the source of roughly one-fifth of the country's CO_2.

Australia
With a scattered population and limited public transport, Australia has one of the highest levels of car ownership in the world.

72% OF GLOBAL TRANSPORT EMISSIONS ARE FROM **ROAD VEHICLES**

29% OF **US** GREENHOUSE GAS EMISSIONS ARE FROM **TRANSPORT**

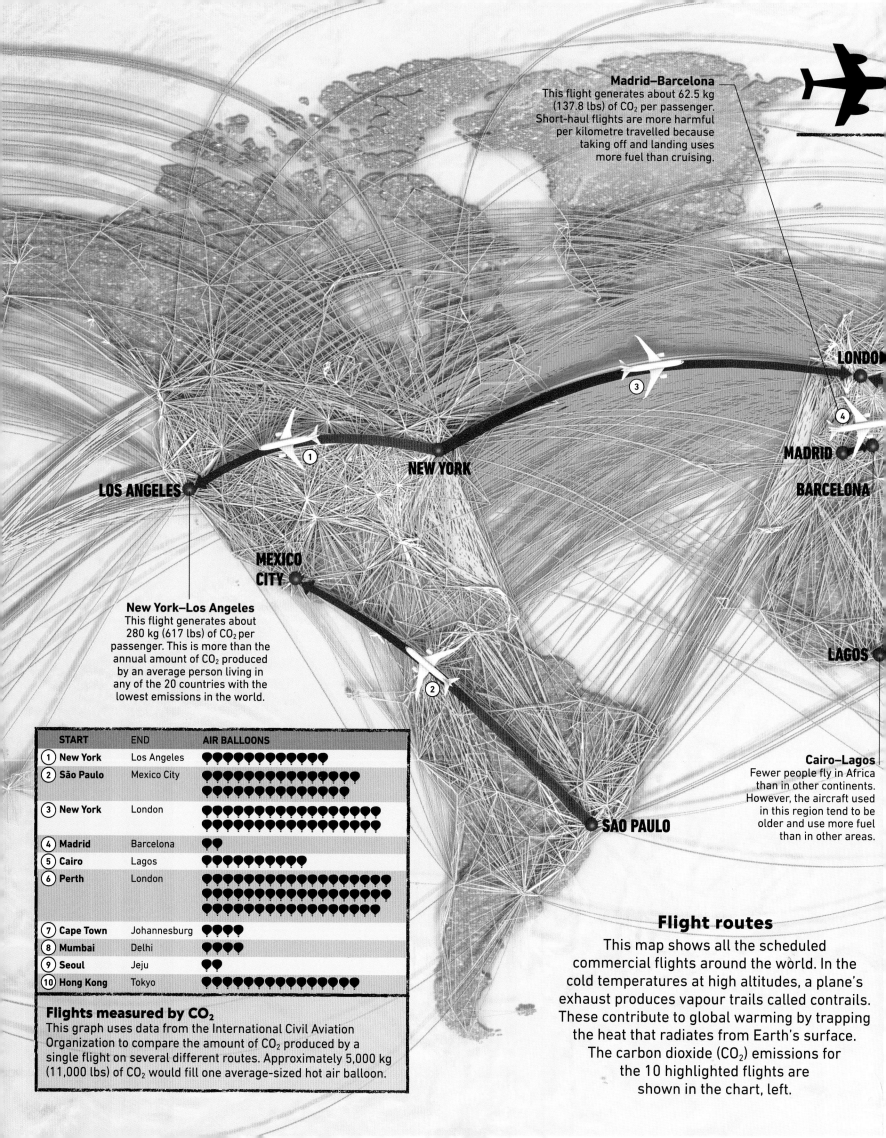

Madrid–Barcelona
This flight generates about 62.5 kg (137.8 lbs) of CO$_2$ per passenger. Short-haul flights are more harmful per kilometre travelled because taking off and landing uses more fuel than cruising.

LONDON

MADRID

BARCELONA

LAGOS

Cairo–Lagos
Fewer people fly in Africa than in other continents. However, the aircraft used in this region tend to be older and use more fuel than in other areas.

NEW YORK

LOS ANGELES

MEXICO CITY

New York–Los Angeles
This flight generates about 280 kg (617 lbs) of CO$_2$ per passenger. This is more than the annual amount of CO$_2$ produced by an average person living in any of the 20 countries with the lowest emissions in the world.

SAO PAULO

	START	END	AIR BALLOONS
①	New York	Los Angeles	🎈🎈🎈🎈🎈🎈🎈🎈🎈🎈🎈
②	São Paulo	Mexico City	🎈🎈🎈🎈🎈🎈🎈🎈🎈🎈🎈🎈🎈🎈🎈🎈🎈🎈🎈🎈🎈🎈🎈🎈🎈🎈🎈🎈
③	New York	London	🎈🎈🎈🎈🎈🎈🎈🎈🎈🎈🎈🎈🎈🎈🎈🎈🎈🎈🎈🎈🎈🎈
④	Madrid	Barcelona	🎈🎈
⑤	Cairo	Lagos	🎈🎈🎈🎈🎈🎈🎈🎈🎈🎈🎈
⑥	Perth	London	🎈🎈
⑦	Cape Town	Johannesburg	🎈🎈🎈🎈
⑧	Mumbai	Delhi	🎈🎈🎈🎈
⑨	Seoul	Jeju	🎈🎈
⑩	Hong Kong	Tokyo	🎈🎈🎈🎈🎈🎈🎈🎈🎈🎈🎈🎈🎈

Flights measured by CO$_2$
This graph uses data from the International Civil Aviation Organization to compare the amount of CO$_2$ produced by a single flight on several different routes. Approximately 5,000 kg (11,000 lbs) of CO$_2$ would fill one average-sized hot air balloon.

Flight routes

This map shows all the scheduled commercial flights around the world. In the cold temperatures at high altitudes, a plane's exhaust produces vapour trails called contrails. These contribute to global warming by trapping the heat that radiates from Earth's surface. The carbon dioxide (CO$_2$) emissions for the 10 highlighted flights are shown in the chart, left.

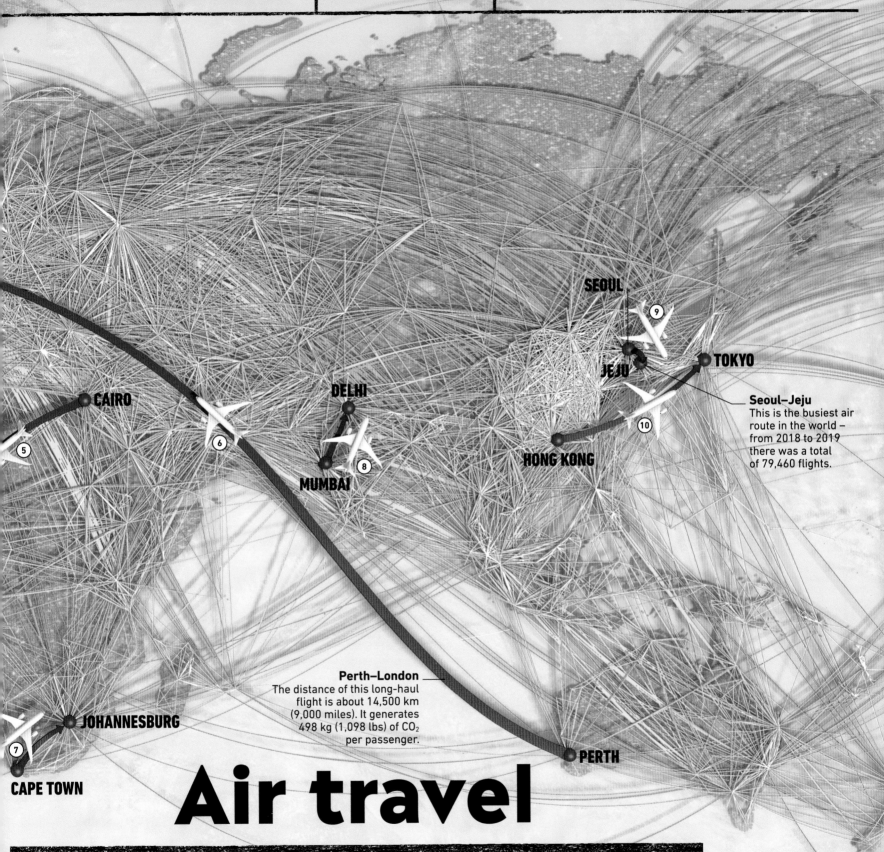

BILLION PASSENGERS FLEW AND NEARLY **38 MILLION** **FLIGHTS** TOOK PLACE **IN 2018**

IN 2019 ALONE, **AVIATION** PRODUCED **915** **MILLION TONNES** OF CO_2

BETWEEN **2013** AND **2018**, A RISE IN THE NUMBER OF **FLIGHTS** INCREASED **GREENHOUSE GAS EMISSIONS** BY **32%**

SEOUL

CAIRO

DELHI

MUMBAI

JEJU

TOKYO

Seoul–Jeju
This is the busiest air route in the world – from 2018 to 2019 there was a total of 79,460 flights.

HONG KONG

Perth–London
The distance of this long-haul flight is about 14,500 km (9,000 miles). It generates 498 kg (1,098 lbs) of CO_2 per passenger.

JOHANNESBURG

PERTH

Air travel

CAPE TOWN

As global prosperity has increased, greater numbers of people can afford to fly commercially for business and pleasure. Today, aviation produces around two per cent of the world's greenhouse gas emissions.

6. Distribution and retail

Jeans are transported by road from warehouses to retail stores and as online deliveries to homes. Transport emits GHGs from diesel combustion, and shops use a lot of energy.

 **10%** OF ALL **GHGs** ARE PRODUCED BY THE **FASHION INDUSTRY**

7. Customer use

The carbon footprint of making a trip to the shops to buy clothes is easily overlooked, as is the high cost to the environment of washing, drying, and ironing them over their lifetime.

FINISH

0.1 KG GHGs

1.7 KG GHGs

0.5 KG GHGs

= TOTAL 11 KG GHGs

8. Garment disposal

Most clothes end up as waste, with harmful effects on the climate. The average consumer buys a new pair of jeans every year and wears them 200 times, but wears a dress just 10 times.

5. Jeans import

Once made, jeans are shipped around the world for sale. Most jeans sold in the USA today are manufactured in China, Mexico, or Bangladesh, where labour is cheap.

Jeans life cycle

A pair of jeans contributes 11 kg of GHGs in its lifetime. This map shows its journey around the world.

START

Fast fashion

1.4 KG GHGs

The fashion industry produces a vast amount of greenhouse gases (GHGs). Trends are constantly updated and clothes are cheap, encouraging consumers to fill their wardrobes with garments that may soon be discarded but will leave a carbon footprint for years to come.

1. Cotton growing

Intensive farming of cotton produces greenhouse gas emissions, mainly nitrous oxide from fertilizer use. Growing cotton also uses a lot of water – the amount required to make one pair of jeans is as much as a person drinks in 10 years.

85% OF USED CLOTHES END UP IN **LANDFILL** OR ARE **INCINERATED**

 A TYPICAL UK **HOUSEHOLD PRODUCES 15** TONNES OF **GHGs A YEAR** FROM ITS **CLOTHING**

KEY

The map shows the greenhouse gases produced at each stage of the life cycle of a pair of jeans.

● Step in the production of jeans

--- Transport of materials or product

4. Garment making

Cutting, sewing, washing, drying, ironing, and packaging jeans uses energy from burning fossil fuels, which emits GHGs. Special finishes such as stone-washing use more energy and water. Up to 20 per cent of fabric is wasted in cutting and dumped or burned.

1.8 KG GHGs

CONTINUES ➤

5.4 KG GHGs

3. Fabric production

Spinning cotton yarn, then dyeing it and weaving it into denim with elastane, a synthetic fibre made from fossil fuels, emits more greenhouse gases than any other part of the process. It also uses and pollutes a lot of water.

 0.1 KG GHGs

2. Cotton export

Shipping raw cotton from the plantation where it is grown to the factories where it will be processed produces relatively few emissions, adding little to its carbon footprint.

Textile waste

These bales of used clothing are ready to be sorted at a warehouse in Senegal. Around 15 per cent of clothes are reused secondhand, upcycled into new items, or recycled globally. The rest end up in landfill or are incinerated. Natural fibres such as cotton biodegrade, but synthetic fabrics such as polyester do not decompose easily, and shed microfibres into water. Burning materials releases GHGs and toxic pollutants into the atmosphere.

The impacts of climate change

How does climate change affect the planet?

The climate emergency has had an impact on all parts of the globe, from scorched deserts and frozen poles to clouds in the atmosphere and the depths of the ocean. As these environments change, so do the lives of people and other living things all around the planet.

EFFECTS ON HUMANS AND HABITATS

Habitat loss and extinctions
Natural habitats across the world are being threatened by climate change, and biodiversity is at risk as animals struggle to adapt.

Climate migration
As the climate changes, people are leaving their homes, and even countries, to escape floods, droughts, and other extremes.

Rising temperatures
Global average temperature is a measure of how the climate is changing. This chart shows by how many degrees Celsius (1°C = 1.8°F) the average temperature has increased – from the average taken before 1900 to that of today.

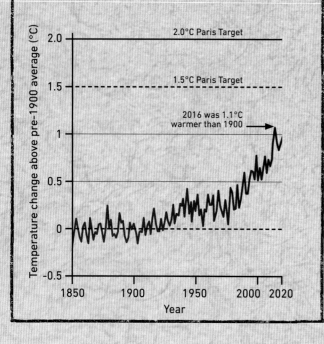

Chart: Temperature change above pre-1900 average (°C) vs Year (1850–2020)
- 2.0°C Paris Target
- 1.5°C Paris Target
- 2016 was 1.1°C warmer than 1900

ENVIRONMENTAL EFFECTS

Changing weather

Changes in the atmosphere cause shifts in rainfall patterns, and increase the likelihood of extreme events such as hurricanes.

Planet under pressure

As the world becomes warmer, the climate is changing in different ways, making lives more challenging.

Heating Earth

Land areas have been exposed to warmer and drier conditions, making them more vulnerable to drought and wildfires.

Warming seas

The oceans are warming and sea levels are rising, changing the lives of animals and people who depend on it.

Melting ice

Melting glaciers contribute to rising sea level, and sea ice is shrinking across the Arctic.

Melting Arctic ice
Rising temperatures are causing Arctic sea ice to melt. In September 2012, the lowest ever ice extent was recorded at 3,410,000 sq km (1,320,000 sq miles).

-1°C

+1°C

Melting glaciers
Glaciers in the European Alps, and in other tall mountain ranges across the world, are melting due to increasing temperatures. Experts think that half of Alpine ice will be gone by 2050.

Global warming

The direct effect of the increase in greenhouse gases in Earth's atmosphere is a rise in temperatures. Global warming is causing droughts, the melting of polar ice, and the heating of the oceans. Rising temperatures also generate extreme weather events.

+1.5°C

Heating oceans
The oceans absorb more than 90 per cent of the heat trapped by greenhouse gases. A rise in ocean temperature has reached record levels over the past decade.

Cooling spots
Regions of the Southern Ocean have become cooler. Scientists believe this may be due to the effects that global warming is having on atmospheric and ocean currents.

TEMPERATURE KEY
The map shows regions that were hotter in 2013–2017 than they were in the period 1950–1980 shaded in red. Changes in the climate are making some areas cooler, and these are shown in blue.

Maximum increase
4°C (7.2°F)

°Celsius
-2   -2   0   1   2

-3   -2   -1   0   1   2   3
°Fahrenheit

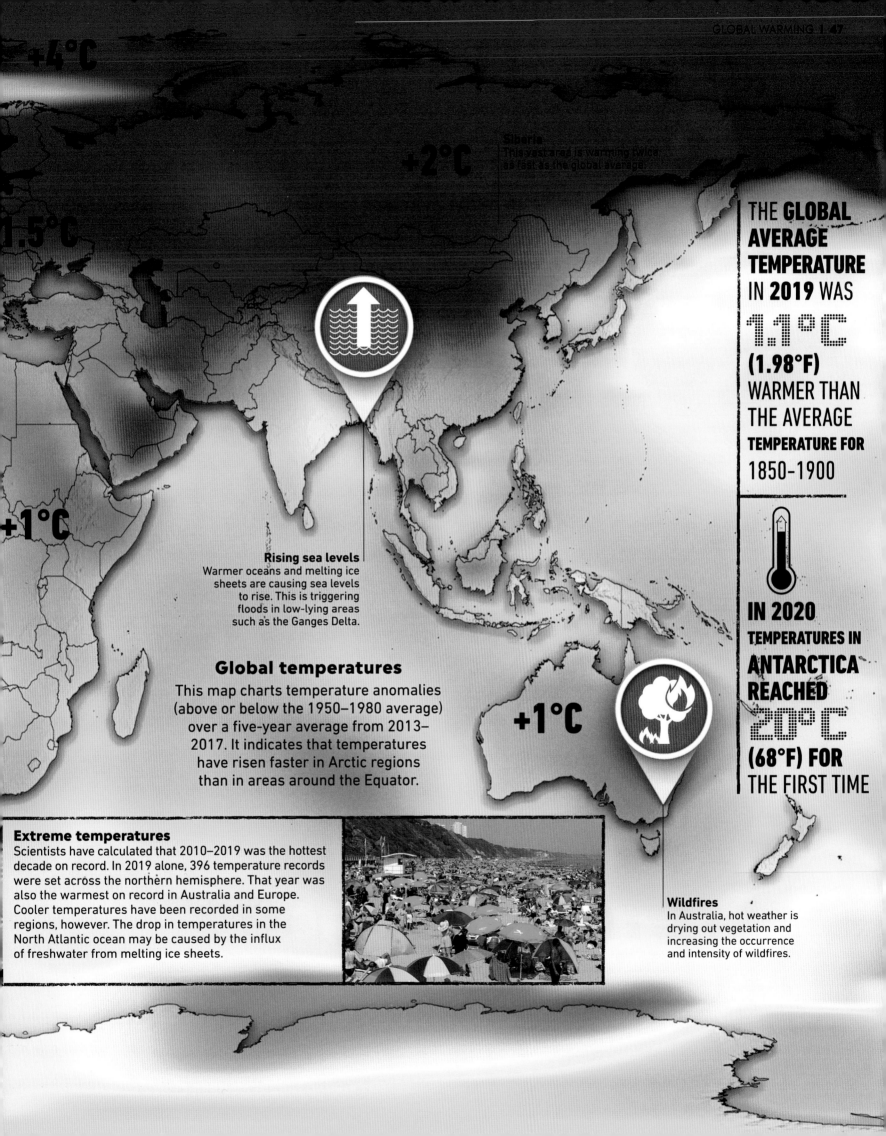

+4°C

+2°C

Siberia
This vast area is warming twice
as fast as the global average.

1.5°C

+1°C

THE **GLOBAL
AVERAGE
TEMPERATURE**
IN **2019** WAS

1.1°C
(1.98°F)
WARMER THAN
THE AVERAGE
TEMPERATURE FOR
1850–1900

IN **2020**
**TEMPERATURES IN
ANTARCTICA
REACHED
20°C**
(68°F) FOR
THE FIRST TIME

Rising sea levels
Warmer oceans and melting ice
sheets are causing sea levels
to rise. This is triggering
floods in low-lying areas
such as the Ganges Delta.

Global temperatures

This map charts temperature anomalies
(above or below the 1950–1980 average)
over a five-year average from 2013–
2017. It indicates that temperatures
have risen faster in Arctic regions
than in areas around the Equator.

+1°C

Extreme temperatures
Scientists have calculated that 2010–2019 was the hottest
decade on record. In 2019 alone, 396 temperature records
were set across the northern hemisphere. That year was
also the warmest on record in Australia and Europe.
Cooler temperatures have been recorded in some
regions, however. The drop in temperatures in the
North Atlantic ocean may be caused by the influx
of freshwater from melting ice sheets.

Wildfires
In Australia, hot weather is
drying out vegetation and
increasing the occurrence
and intensity of wildfires.

California, USA, 2018
The most destructive Californian wildfires on record, in which 22,000 structures were destroyed, and 95 lives were lost.

Canada, 2019
A temperature of 21°C (69.8°F) was recorded at the world's most northerly inhabited spot for the first time.

Greenland, 2019
11.5 billion tonnes (12.5 billion tons) of ice melted from an ice sheet in just one day.

Alaska, USA, 2019
Sea ice melted during the warmest year on record. A temperature of 32.2°C (90°F) was registered for the first time.

UK, 2020
Storms brought heavy rainfall and floods.

France, 2019
A heatwave drove temperatures to a record-breaking 45.9°C (114.6°F).

Texas, USA, 2017
During Hurricane Harvey, record rainfall over 4 days led to catastrophic flooding.

Canada, 2020
A record-breaking 76.2 cm (2.5 ft) of snow fell in one day in Newfoundland.

Hurricane Maria, 2017
This deadly storm caused major destruction in the Caribbean.

Spain, 2019
A heatwave triggered the biggest wildfires in 20 years.

Cuba, 2020
The island had its hottest day on record – 39.3°C (102.7°F).

Peru, 2017
Extreme rainfall triggered mudslides and burst riverbanks. Scientists believe human activity was a partial cause.

Hurricane Irma, 2017
The most powerful Atlantic hurricane in a decade, Irma killed more than 134 people.

West Africa, 2012
Floods destroyed homes and crops.

Bolivia and Paraguay, 2017
Major landslides and floods were caused by heavy rainfall.

Southern Africa, 2019
Severe drought wiped out livestock and crops.

Chile, 2019
Temperatures reached 32.2°C (90°F) in the far south of the country for the first time.

Argentina, 2019
A record 224 mm (8.8 in) of rain fell in one day.

Wild weather worldwide

This map shows some of the most extreme weather events that have occured over the past decade, including record-breaking temperatures in many countries. Climate experts have shown that human activity has made these events more likely to happen.

Chile, 2017
High temperatures, drought and strong winds led to Chile's worst wildfires in recent history.

Extreme weather

In recent decades, extreme weather events, such as searing heatwaves, flash floods, and huge, powerful hurricanes, have been striking with greater frequency all over the world. Their sheer number, intensity, and distribution indicate a major change in the Earth's climate.

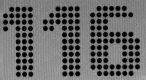 **116** **HEATWAVES** SINCE THE YEAR 2000 HAVE BEEN **MORE SEVERE** AS A RESULT OF **CLIMATE CHANGE**

THE **COST** OF **EXTREME WEATHER** EVENTS WAS MORE THAN **$100 BILLION** IN **2019**

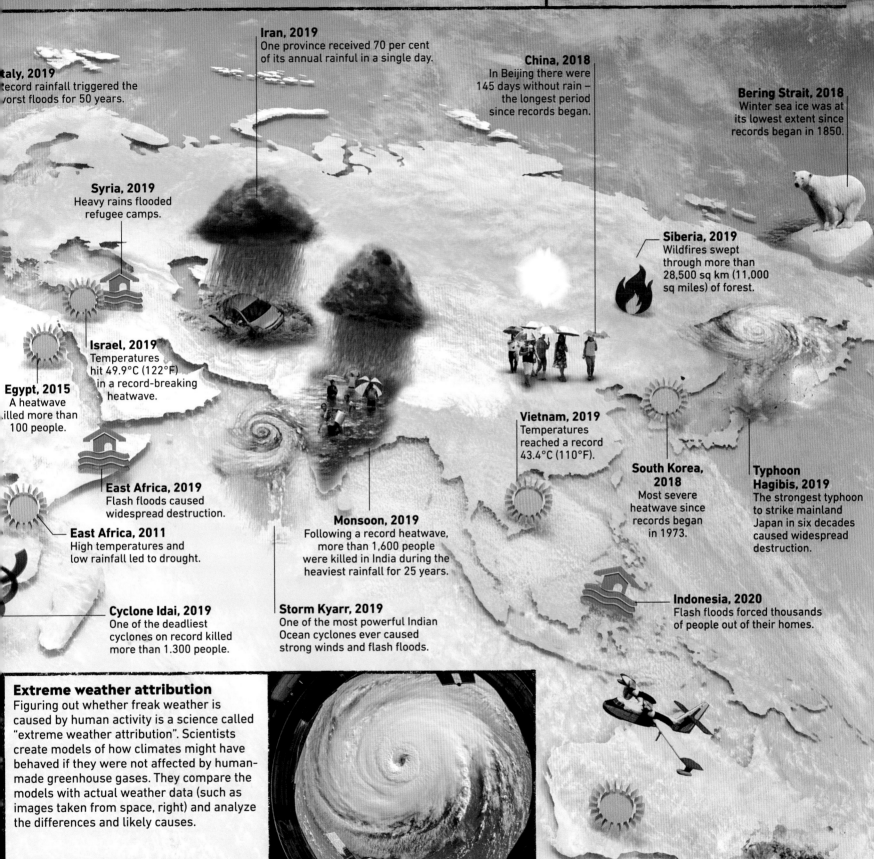

Iran, 2019
One province received 70 per cent of its annual rainful in a single day.

China, 2018
In Beijing there were 145 days without rain – the longest period since records began.

Bering Strait, 2018
Winter sea ice was at its lowest extent since records began in 1850.

Italy, 2019
Record rainfall triggered the worst floods for 50 years.

Syria, 2019
Heavy rains flooded refugee camps.

Siberia, 2019
Wildfires swept through more than 28,500 sq km (11,000 sq miles) of forest.

Israel, 2019
Temperatures hit 49.9°C (122°F) in a record-breaking heatwave.

Egypt, 2015
A heatwave killed more than 100 people.

Vietnam, 2019
Temperatures reached a record 43.4°C (110°F).

East Africa, 2019
Flash floods caused widespread destruction.

South Korea, 2018
Most severe heatwave since records began in 1973.

Typhoon Hagibis, 2019
The strongest typhoon to strike mainland Japan in six decades caused widespread destruction.

East Africa, 2011
High temperatures and low rainfall led to drought.

Monsoon, 2019
Following a record heatwave, more than 1,600 people were killed in India during the heaviest rainfall for 25 years.

Cyclone Idai, 2019
One of the deadliest cyclones on record killed more than 1.300 people.

Storm Kyarr, 2019
One of the most powerful Indian Ocean cyclones ever caused strong winds and flash floods.

Indonesia, 2020
Flash floods forced thousands of people out of their homes.

Extreme weather attribution

Figuring out whether freak weather is caused by human activity is a science called "extreme weather attribution". Scientists create models of how climates might have behaved if they were not affected by human-made greenhouse gases. They compare the models with actual weather data (such as images taken from space, right) and analyze the differences and likely causes.

Australia, 2020
A record-breaking heatwave fuelled massive wildfires. burning an area of 186,000 sq km (72,000 sq miles).

 **1500** PEOPLE IN **FRANCE** LOST THEIR **LIVES** AS A RESULT OF THE **2019 EUROPEAN HEATWAVE**

SUMMER ARCTIC ICE HAS REDUCED BY ABOUT **40 PER CENT** SINCE **SATELLITE** RECORDS BEGAN IN 1979

60 M (200 FT) IS THE HEIGHT THAT **SEA LEVELS** WOULD **RISE** IF ALL THE **ANTARCTIC ICE SHEETS** MELTED

Melting accelerates
By 2000, the annual minimum extent of the ice had reduced to 6.4 million sq km (2.5 million sq miles). The shrinking of the ice began to accelerate in the 1990s as the impact of greenhouse gases on the climate intensified.

A wider extent
The annual minimum extent of the ice in 1980 was about 7.9 million sq km (3 million sq miles). The majority of ice in 1980 was also thicker than it is today and lasted longer than one year.

Arctic ice extent
This map shows the shrink in the annual minimum extent of ice at the Arctic in 1980, 2000, and 2019. Arctic ice increases in the winter and melts during the summer, reaching its minimum extent in September.

2019

2000

1980

KEY
The shaded areas show Arctic ice extent at different stages since 1980.

◾ 1980
◾ 2000
◻ 2019

Shrinking polar ice

The Arctic, which is made up of frozen sea, is warming faster than anywhere else on the planet. The melting of its sea ice is endangering wildlife and harming the livelihoods of people who rely on polar resources. At the Antarctic, vast swaths of ice situated on land are also disappearing, causing rising sea levels.

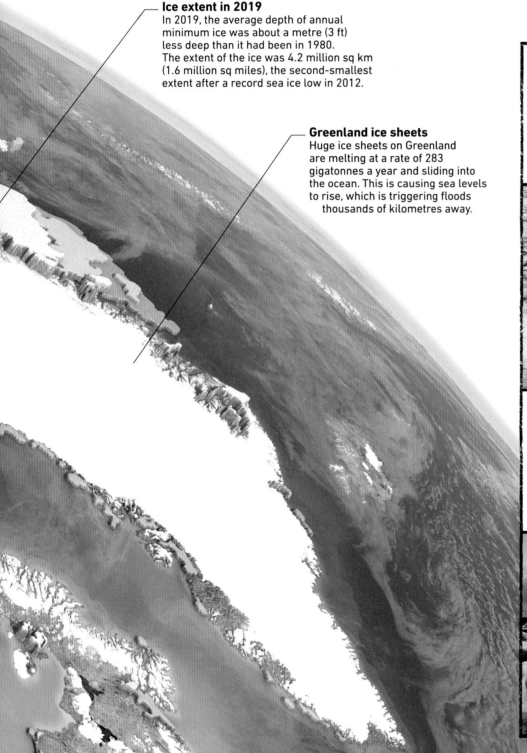

Ice extent in 2019
In 2019, the average depth of annual minimum ice was about a metre (3 ft) less deep than it had been in 1980. The extent of the ice was 4.2 million sq km (1.6 million sq miles), the second-smallest extent after a record sea ice low in 2012.

Greenland ice sheets
Huge ice sheets on Greenland are melting at a rate of 283 gigatonnes a year and sliding into the ocean. This is causing sea levels to rise, which is triggering floods thousands of kilometres away.

Arctic amplification
The fast rate of warming at the Arctic is partly due to the interaction between sunlight, ice, and the ocean. The white sea ice reflects sunlight, but when it melts it uncovers the dark ocean (as seen below), which absorbs a larger proportion of the Sun's energy. This warms the sea, amplifying the heating effect.

Antarctica
Like the Arctic, the Antarctic is warming, but sea ice in nearby oceans is increasing rather than melting. Scientists think this is due to changes in wind and ocean currents. However, most Antarctic ice is on the land – in the form of ice sheets – and this is decreasing at a rate of 145 gigatonnes every year.

On the day this photo was taken, more than two billion tonnes of ice melted in Greenland.

Melting ice sheets

Huskies pulling a sledge carrying climatologists from the Danish Meteorological Institute run through ankle-deep water after unusually high temperatures melted part of the solid ice sheet in northern Greenland's Inglefield Bredning fjord. The thick floating sea ice forms in the fjord each year during the winter and melts away in the warmer summer months of July and August. However, above average temperatures in Greenland and the whole Arctic region in 2019 led to the ice melting much earlier than usual. On 12 June, the day before this photograph was taken, the temperature was more than 22°C (40°F) above normal.

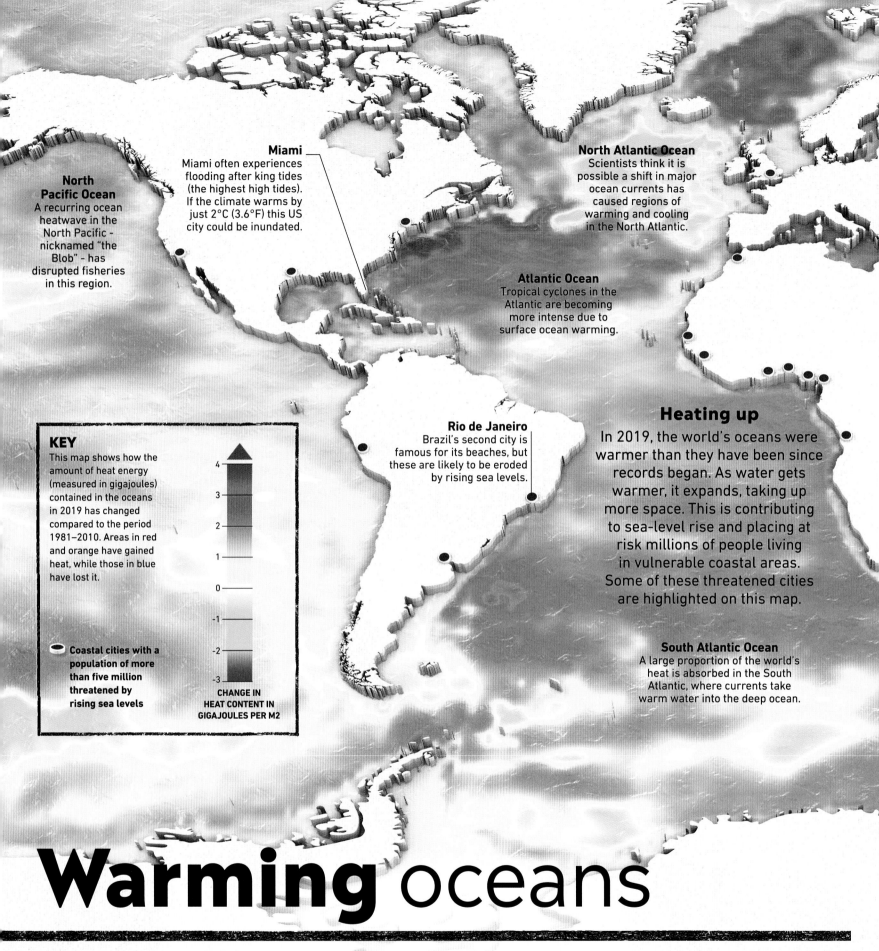

North Pacific Ocean
A recurring ocean heatwave in the North Pacific - nicknamed "the Blob" - has disrupted fisheries in this region.

Miami
Miami often experiences flooding after king tides (the highest high tides). If the climate warms by just 2°C (3.6°F) this US city could be inundated.

North Atlantic Ocean
Scientists think it is possible a shift in major ocean currents has caused regions of warming and cooling in the North Atlantic.

Atlantic Ocean
Tropical cyclones in the Atlantic are becoming more intense due to surface ocean warming.

Rio de Janeiro
Brazil's second city is famous for its beaches, but these are likely to be eroded by rising sea levels.

Heating up
In 2019, the world's oceans were warmer than they have been since records began. As water gets warmer, it expands, taking up more space. This is contributing to sea-level rise and placing at risk millions of people living in vulnerable coastal areas. Some of these threatened cities are highlighted on this map.

South Atlantic Ocean
A large proportion of the world's heat is absorbed in the South Atlantic, where currents take warm water into the deep ocean.

KEY
This map shows how the amount of heat energy (measured in gigajoules) contained in the oceans in 2019 has changed compared to the period 1981–2010. Areas in red and orange have gained heat, while those in blue have lost it.

4
3
2
1
0
-1
-2
-3

Coastal cities with a population of more than five million threatened by rising sea levels

CHANGE IN HEAT CONTENT IN GIGAJOULES PER M2

Warming oceans

As levels of greenhouse gases in the atmosphere rise and the planet warms, 90 per cent of this excess heat is being absorbed by the world's seas. The oceans can absorb a large amount of energy with only a small increase in temperature, but this warming is causing sea levels to rise and damaging delicate ocean ecosystems.

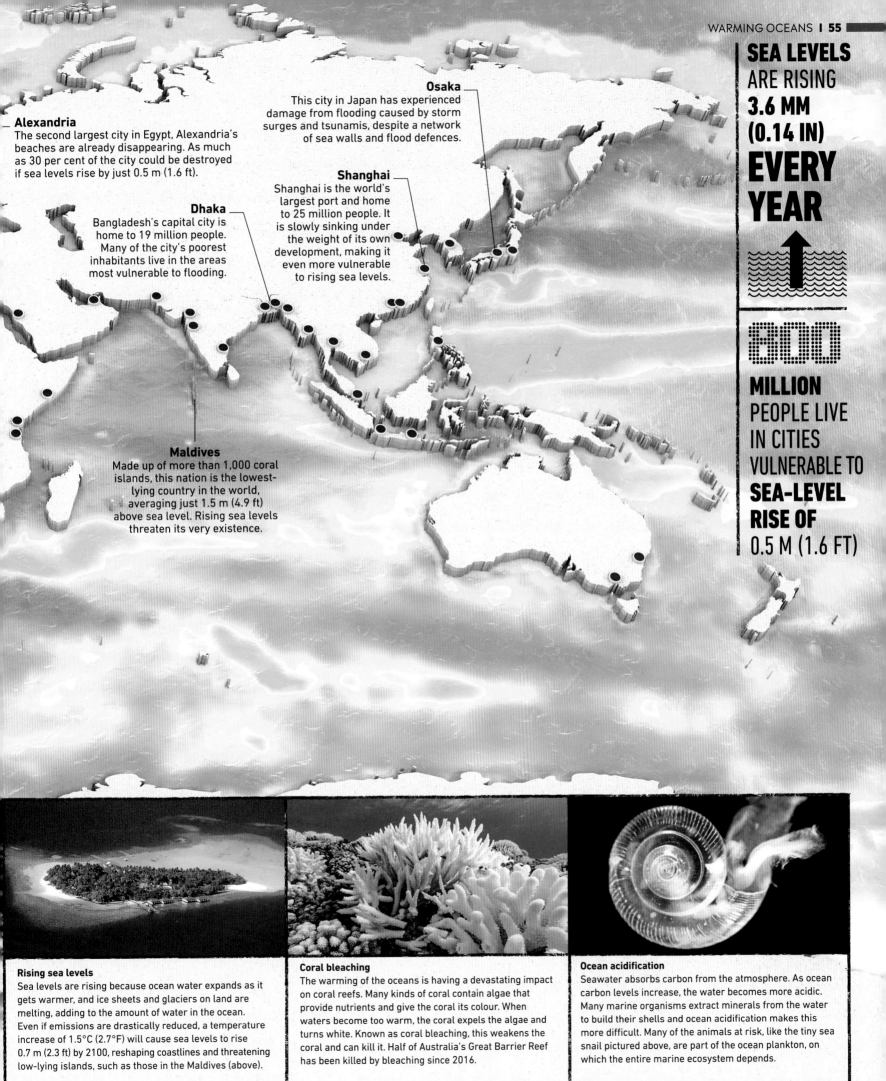

Alexandria
The second largest city in Egypt, Alexandria's beaches are already disappearing. As much as 30 per cent of the city could be destroyed if sea levels rise by just 0.5 m (1.6 ft).

Osaka
This city in Japan has experienced damage from flooding caused by storm surges and tsunamis, despite a network of sea walls and flood defences.

Shanghai
Shanghai is the world's largest port and home to 25 million people. It is slowly sinking under the weight of its own development, making it even more vulnerable to rising sea levels.

Dhaka
Bangladesh's capital city is home to 19 million people. Many of the city's poorest inhabitants live in the areas most vulnerable to flooding.

Maldives
Made up of more than 1,000 coral islands, this nation is the lowest-lying country in the world, averaging just 1.5 m (4.9 ft) above sea level. Rising sea levels threaten its very existence.

SEA LEVELS ARE RISING 3.6 MM (0.14 IN) EVERY YEAR

600 MILLION PEOPLE LIVE IN CITIES VULNERABLE TO SEA-LEVEL RISE OF 0.5 M (1.6 FT)

Rising sea levels
Sea levels are rising because ocean water expands as it gets warmer, and ice sheets and glaciers on land are melting, adding to the amount of water in the ocean. Even if emissions are drastically reduced, a temperature increase of 1.5°C (2.7°F) will cause sea levels to rise 0.7 m (2.3 ft) by 2100, reshaping coastlines and threatening low-lying islands, such as those in the Maldives (above).

Coral bleaching
The warming of the oceans is having a devastating impact on coral reefs. Many kinds of coral contain algae that provide nutrients and give the coral its colour. When waters become too warm, the coral expels the algae and turns white. Known as coral bleaching, this weakens the coral and can kill it. Half of Australia's Great Barrier Reef has been killed by bleaching since 2016.

Ocean acidification
Seawater absorbs carbon from the atmosphere. As ocean carbon levels increase, the water becomes more acidic. Many marine organisms extract minerals from the water to build their shells and ocean acidification makes this more difficult. Many of the animals at risk, like the tiny sea snail pictured above, are part of the ocean plankton, on which the entire marine ecosystem depends.

Precipitation patterns

This map shows the difference in average rainfall between the periods 1980–2000 and 2001–2020. In general, dry areas are getting drier and wet areas are becoming wetter, and rainstorms are increasingly intense.

Drought, sub-Sahara
Lower rainfall in sub-Saharan Africa is affecting crop yields and causing water scarcity in a region that has a fast-growing population.

Atlantic Ocean
Warmer temperatures are increasing the amount of water vapour in the atmosphere and fuelling storms and hurricanes over the ocean.

Drought, USA
Heatwaves and low rainfall led to drought in many areas of the USA during 2010–2013, affecting crop yields and food prices.

Heavy rainfall, South America
Unusually intense downpours causing flash floods and landslides are becoming more common in tropical countries, such as Colombia, Brazil, and Peru.

Torrential rain, West Africa
In Kinshasa, Democratic Republic of Congo, flash floods and landslides destroyed bridges, roads, and homes in 2019.

Radical
rainfall

Climate change has altered rainfall patterns across the world. While some areas are seeing less rainfall, others suffer from sudden floods. These changes have occurred rapidly in the last 20 to 40 years, providing little time for communities to adapt.

Desertification
The reduction in rainfall is making some fertile land more arid through a process called desertification. Drylands, such as those in East Africa (above), form about 40 per cent of Earth's land, and this is set to increase. Most drylands are located in developing countries that rely heavily on rain-fed agriculture.

$44 BILLION – THE GLOBAL ANNUAL COST OF DROUGHTS WORLDWIDE

Drought, Caucasus
Rising temperatures and a reduction in rainfall are eroding once fertile soil and causing crop disease in the region between the Black and Caspian seas.

Drought, Siberia
Low rainfall is increasing drought conditions in Siberia. Dry vegetation is contributing to forest fires here.

Extreme rain, India
A threefold increase in extreme rain events in central India between 1950 and 2015 has led to widespread flooding.

Rainfall key
Red shaded areas indicate where rainfall is lower. Blue shaded areas indicate where rainfall is higher.

CHANGE IN AVERAGE DAILY PRECIPITATION
2001–2020 minus 1980–2000

Highest rainfall increase
18.7 mm (0.74 in)

Millimetres		Inches
2		0.08
1		0.04
0		0
-1		-0.04
-2		-0.08

Highest rainfall decrease
-5.9 mm (-0.23 in)

On some days the increase and decrease in rainfall was even greater than that shown on the map.

Floods, Kenya
In Kenya in 2018, 85 sq km (33 sq miles) of farmland were destroyed by floods following a severe drought.

Drought, Cambodia and Vietnam
Rising temperatures and low rainfall across this region are causing droughts to be more frequent.

Downpours, Indonesia
Monsoon rainfall and typhoons are becoming more intense and harder to predict.

Drought, Australia
Australia has seen years of reduced rainfall. In 2019, the country suffered its worst drought since records began in 1900.

Devastating downpours
Global warming is causing intense spates of rain, particularly in the tropics. This is because warmer air holds more moisture, creating a build-up of water vapour that leads to rain. Flooding triggered by torrential downpours not only destroys homes and roads, but causes crop failure and soil erosion (shown, right, in India). It is making access to food more difficult in many communities where food is already scarce.

 MOST PARTS OF THE **WORLD** WILL HAVE A 16-24 PER CENT **INCREASE** IN **HEAVY RAIN INTENSITY** BY **2100**

 **BILLION** PEOPLE LIVE IN REGIONS WHERE **WATER DEMAND** EXCEEDS SUPPLY

34
PEOPLE WERE KILLED BY THE 2019-2020 FIRES

6,000 HOMES WERE DESTROYED

A record-breaking heatwave following a year of record low rainfall in 2019 led to Australia's worst ever drought. Temperatures were at least 1.5°C (2.7°F) higher than average, and just 278 mm (11 in) of rain fell – 40 per cent less than in a typical year. There is overwhelming evidence that the drought, which created a tinderbox for wildfires, was a result of climate change.

Uncontrollable fires destroyed homes and killed people across Australia. The fires blazed most ferociously in the populous southeast, where they came close to the capital, Canberra (above). Choking clouds of smoke engulfed cities, and hundreds of smaller settlements were evacuated. Firefighters were unable to tackle the raging fires due to the incredible heat of the blazes.

Australian wildlife, such as kangaroos and koalas, made the news around the world as their habitat was destroyed by fire. They were just some of the hundreds of millions of animals, including livestock, that died in the fires. Those that survived suffered a catastrophic loss of food and water, leading to severe malnutrition.

Indian Ocean
Changes to the system of Indian Ocean currents that circulate warmer and cooler waters between Australia and India have contributed to Australia's drier summers.

WESTERN AUSTRALIA

Perth

Albany

Western Australia
Despite a severe tropical cyclone that caused flooding early in 2019, the state suffered the warmest and one of the driest years on record. Fires raged through forests after rain-bearing winds shifted south.

Raging inferno
After the hottest, driest year on record, bushfires raged across Australia from late 2019, eventually burning 100,000 sq km (38,500 sq miles) – an area the size of Iceland – and billowing a toxic cloud of smoke as large as Europe across the world. More than 100 separate fires burned over a three-month period, producing intense heat and releasing more carbon dioxide than Australia usually produces in an entire year.

Australian bushfires

Bushfires are a part of life in Australia, but unprecedented blazes swept the country in 2020, destroying more than 20 per cent of native forest. Climate scientists have attributed the ferocity of the fires to hotter, drier weather caused by climate change.

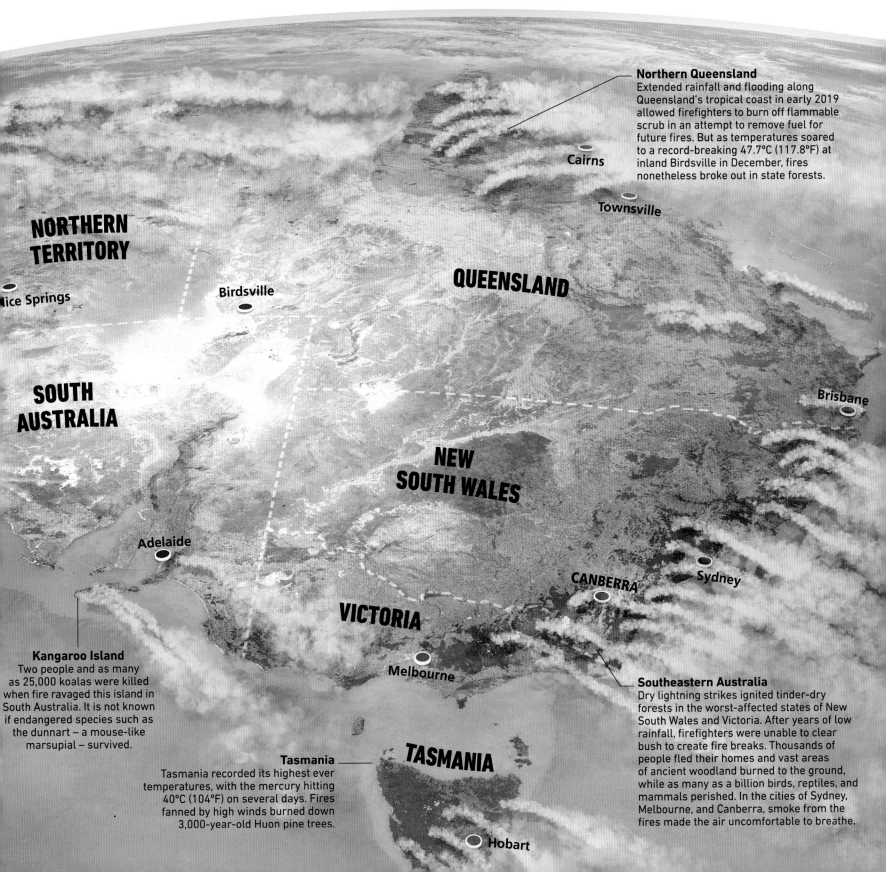

Northern Queensland
Extended rainfall and flooding along Queensland's tropical coast in early 2019 allowed firefighters to burn off flammable scrub in an attempt to remove fuel for future fires. But as temperatures soared to a record-breaking 47.7°C (117.8°F) at inland Birdsville in December, fires nonetheless broke out in state forests.

NORTHERN TERRITORY

Alice Springs

Birdsville

QUEENSLAND

SOUTH AUSTRALIA

Cairns

Townsville

Brisbane

NEW SOUTH WALES

Adelaide

CANBERRA

Sydney

VICTORIA

Melbourne

Kangaroo Island
Two people and as many as 25,000 koalas were killed when fire ravaged this island in South Australia. It is not known if endangered species such as the dunnart – a mouse-like marsupial – survived.

TASMANIA

Tasmania
Tasmania recorded its highest ever temperatures, with the mercury hitting 40°C (104°F) on several days. Fires fanned by high winds burned down 3,000-year-old Huon pine trees.

Hobart

Southeastern Australia
Dry lightning strikes ignited tinder-dry forests in the worst-affected states of New South Wales and Victoria. After years of low rainfall, firefighters were unable to clear bush to create fire breaks. Thousands of people fled their homes and vast areas of ancient woodland burned to the ground, while as many as a billion birds, reptiles, and mammals perished. In the cities of Sydney, Melbourne, and Canberra, smoke from the fires made the air uncomfortable to breathe.

Threat to biodiversity

This map shows where biodiversity is most under threat from the effects of climate change. A staggering 75 per cent of the land-based environment has been severely altered by human activity, with disastrous effects for plants and animals.

25% OF KNOWN **SPECIES** ARE IN **DANGER** OF **DISAPPEARING**

1 MILLION ANIMAL AND PLANT SPECIES ARE AT RISK OF **EXTINCTION**

Frigid bumblebee
This bumblebee, which lives in the cool, coniferous forests of North America, survives in a narrow temperature range.

American pika
This mountain creature is affected by lack of the snow that keeps it warm in winter, and by hotter summers. Forced to move higher, where it is cooler but harsher, it struggles to survive.

Monarch butterfly
Climate change and habitat loss are threatening monarch butterflies' migration as summer heat drives them further north to breed and farming kills off the milkweed their caterpillars eat.

Polar bear
The Arctic is warming twice as much as the global average, causing sea ice to melt. Polar bears eat seals, which they hunt on the ice. Seals also need ice to raise their young.

Loggerhead turtle
Increasing sand temperatures on the beaches where marine turtles nest are causing more females than males to hatch.

Dyeing poison frog
These brightly coloured frogs cannot regulate their body temperature, so they find it hard to survive in a warmer climate when cool wetland forests are cleared.

African elephant
Already threatened by poaching, habitat loss, and conflict with people for land and food, elephants need large amounts of water to survive, migrate, and reproduce.

Hyacinth macaw
Cattle-ranching and drought threaten the forest habitat and nut diet of these eye-catching parrots, along with myriad other species that together make up the complex Amazon ecosystem.

Antarctic krill
Warmer seas are changing the range and breeding season of krill, the tiny shrimp-like creatures that are the main food source of humpback whales.

Adélie penguin
Numbers of this penguin have declined by 80 per cent since the 1970s and are predicted to decline further as rising sea temperatures affect northern Antarctic communities.

Animals in danger

Climate change and extreme weather are having a severe impact on our planet's natural habitats and biodiversity, from the shrinking Arctic sea ice where polar bears live to raging wildfires that destroy forest homes. The changing climate is also forcing people and animals to compete for dwindling resources.

Arctic fox
Long hunted for their pelts, Arctic foxes have thrived since furs went out of fashion, but rising temperatures and red foxes are driving them further north into the tundra.

European bee-eater
The brightly coloured bird has changed its migratory path from Europe to Africa as the warming climate has driven it further north to breed, and wetland stopovers have shrunk.

Siberian tiger
Fewer than 600 Siberian tigers remain. Felling of their Korean pine forest habitat, combined with longer, hotter, and drier wildfire seasons, is putting them at risk of extinction.

KEY
Red areas show where large numbers of species are at risk of extinction as climate change and habitat destruction endanger the web of life in fragile ecosystems.

Areas at risk

Snow leopard
Vulnerable to poaching for their fur, snow leopards are now also disturbed by shepherds moving their flocks into the high plains and mountains of Central Asia as low grasslands dry out.

Giant panda
The bamboo that giant pandas eat dies off every few years. New shoots may not grow fast enough to meet the bears' needs as climate change drives them higher up into the mountains.

Clownfish
The clownfish's coral reef habitat is under severe threat from ocean acidification and warming oceans. These effects of climate change are disrupting the fish's ability to breed.

Black rhino
East Africa's black rhino population has already been decimated by poaching. Now the rhinos are facing extreme drought and a lack of food and water.

Ring-tailed lemur
The ring-tailed lemur's habitat in the dry forests of southern Madagascar is being destroyed for cattle pasture, leaving this primate under threat from drought as the land becomes more arid.

Mountain gorilla
These rare apes are being driven higher up forested slopes by changes in temperature and rainfall as well as human pressure. This affects their food supply and exposes them to new diseases.

Bornean orangutan
Orangutans in Borneo have lost half of their forest home to logging and oil palm plantations in the last 20 years. Forest loss has also disrupted the water supply, increasing drought and fire risk.

Brush-tailed rock wallaby
These small marsupials have been endangered by the ferocious wildfires sweeping across Australia. Wallabies may survive a fire, but are left with no food if the vegetation they eat has been burned.

Ecosystems and habitats under threat

 Tundra and taiga As sea ice melts and permafrost thaws, Arctic species struggle, while subarctic plants and animals grow and move further north.

 Temperate forests Deciduous woodlands now grow nearer the poles, while rainfall variability makes some wetter and others drier, unbalancing their ecosystems.

 Tropical rainforests Rising temperatures, lower rainfall, drier ground, and increasing wildfire risk transform forest ecosystems.

 Mountains Mountain-top species fight to survive in shrinking habitats as temperatures rise, and other species move up the slopes to take their place.

 Grasslands Warmer temperatures disrupt rainfall, making drought more frequent. Animals have to migrate long distances to find different conditions.

 Wetlands Inland freshwater wetlands are drying out, while rising sea levels are flooding coastal wetlands, affecting birds' migration and breeding.

2019 was Australia's hottest and driest year since records began, with temperatures up by 1.5°C (2.7°F) and rainfall at 40% below average.

Parched earth

A farmer drops feed for his sheep in a dry, dusty paddock on his farm in New South Wales, Australia. Drought is a normal part of Australia's climate, but conditions in recent years have been particularly severe. 2019 saw record high temperatures and consistently low rainfall, making that year's drought the worst for more than 100 years. Rivers ran dry and reservoirs reached critically low levels. The dry conditions also played a part in the devastating bushfires of 2019-2020, when fires raged out of control across large parts of the country. New South Wales was the worst hit state, with 50,000 sq km (19,500 sq miles) of land burned and 2,000 homes destroyed.

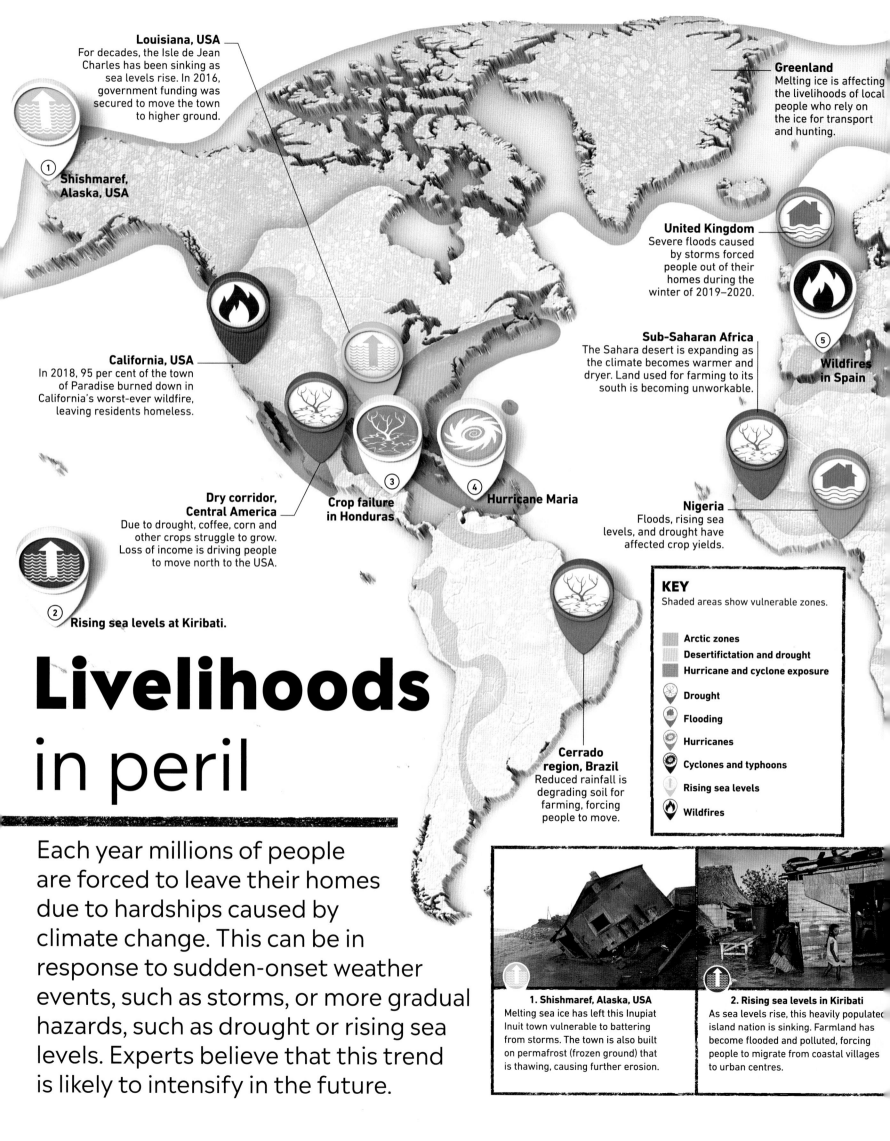

Louisiana, USA
For decades, the Isle de Jean Charles has been sinking as sea levels rise. In 2016, government funding was secured to move the town to higher ground.

Greenland
Melting ice is affecting the livelihoods of local people who rely on the ice for transport and hunting.

① **Shishmaref, Alaska, USA**

United Kingdom
Severe floods caused by storms forced people out of their homes during the winter of 2019–2020.

California, USA
In 2018, 95 per cent of the town of Paradise burned down in California's worst-ever wildfire, leaving residents homeless.

Sub-Saharan Africa
The Sahara desert is expanding as the climate becomes warmer and dryer. Land used for farming to its south is becoming unworkable.

⑤ **Wildfires in Spain**

Dry corridor, Central America
Due to drought, coffee, corn and other crops struggle to grow. Loss of income is driving people to move north to the USA.

③ **Crop failure in Honduras**

④ **Hurricane Maria**

Nigeria
Floods, rising sea levels, and drought have affected crop yields.

② **Rising sea levels at Kiribati.**

Livelihoods in peril

Cerrado region, Brazil
Reduced rainfall is degrading soil for farming, forcing people to move.

KEY
Shaded areas show vulnerable zones.

- Arctic zones
- Desertifictation and drought
- Hurricane and cyclone exposure
- Drought
- Flooding
- Hurricanes
- Cyclones and typhoons
- Rising sea levels
- Wildfires

Each year millions of people are forced to leave their homes due to hardships caused by climate change. This can be in response to sudden-onset weather events, such as storms, or more gradual hazards, such as drought or rising sea levels. Experts believe that this trend is likely to intensify in the future.

1. Shishmaref, Alaska, USA
Melting sea ice has left this Inupiat Inuit town vulnerable to battering from storms. The town is also built on permafrost (frozen ground) that is thawing, causing further erosion.

2. Rising sea levels in Kiribati
As sea levels rise, this heavily populated island nation is sinking. Farmland has become flooded and polluted, forcing people to migrate from coastal villages to urban centres.

Melting permafrost
As frozen soils melt, arable land is becoming unfit for farming so people are moving.

7 MILLION PEOPLE WERE DISPLACED BY WEATHER EVENTS IN THE FIRST HALF OF 2019

IN 2015, 85% OF PEOPLE DISPLACED BY SUDDEN ONSET EVENTS WERE IN ASIA

Syria
From 2007–2010 a severe drought ravaged Syria, Iraq, and Turkey. It pushed more than 1.5 million people from the countryside into the cities.

Siberia
In 2019 wildfires posed a threat to the population.

Afghanistan
Drought triggered migration in 2019.

Shanghai, China
Shanghai has been described as the most vulnerable major city in the world to flooding. HIghly populated and on the coast, it is also has several waterways running through it.

Floods in Bangladesh
⑥

Philippines
In 2013, 6 million people were displaced by Typhoon Haiyan.

Ethiopia,
Droughts have destroyed crops, forcing people to move to cities or displaced person's camps.

Mekong Delta
Sea levels are rising in Vietnam.

Jakarta
As sea levels rise, Indonesia's capital city, Jakarta, is sinking. There are plans to move the capital to a different island.

Global displacement

This map shows areas of the world where populations are particularly vulnerable to the effects of climate change. It also highlights places where climate change has already forced people to move.

Cyclone Idai
Idai caused the displacement of 600,000 people in 2019.

Fiji, Tuvalu, and Samoa
These islands are threatend by rising sea levels.

Southeastern Australia
Hundreds of thousands of people were displaced during huge wildfires fuelled by prolonged drought in 2019–2020.

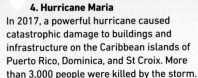

3. Crop failure in Honduras
...treme weather patterns in Honduras ...d other Central American countries are ...fecting coffee yields. Heavy rain fuels a ...ust" fungus (pictured above), while too ...le rain causes the plant to dry up.

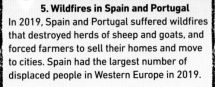

4. Hurricane Maria
In 2017, a powerful hurricane caused catastrophic damage to buildings and infrastructure on the Caribbean islands of Puerto Rico, Dominica, and St Croix. More than 3,000 people were killed by the storm.

5. Wildfires in Spain and Portugal
In 2019, Spain and Portugal suffered wildfires that destroyed herds of sheep and goats, and forced farmers to sell their homes and move to cities. Spain had the largest number of displaced people in Western Europe in 2019.

6. Floods in Bangladesh
Floods caused by storms in villages in the Ganges Delta have triggered mass migration. Salt water has polluted wells and made rice fields barren. Sea levels are also rising in the Ganges Delta.

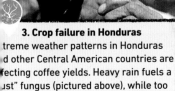

Action on climate change

What can we do about climate change?

To halt climate change, we must rapidly reduce all greenhouse gas emissions. Government initiatives are key, but individual actions play an important role, too. Communities must also adapt to minimize the effects of climate change.

INDIVIDUAL ACTION

Be heard
Tell governments, businesses, and schools how much the issue of climate change matters to us by joining action groups and joining protests.

Low-emission diet
Changing what we eat, for example by avoiding high-emission foods like beef, will help reduce the carbon footprint of agriculture.

Sustainable living
We can change our lifestyles by making more climate-friendly choices, such as buying less, reusing and recycling more, and using low-emission transport.

Our carbon budget
Scientists have worked out how much carbon (in the form of fossil fuels) we can burn between now and 2050 to still keep the average global temperature rise below 2°C (3.6°F) by 2050.

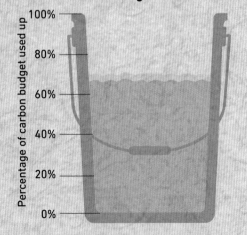

Carbon budget for 2°C limit

Percentage of carbon budget used up

100%
80%
60%
40%
20%
0%

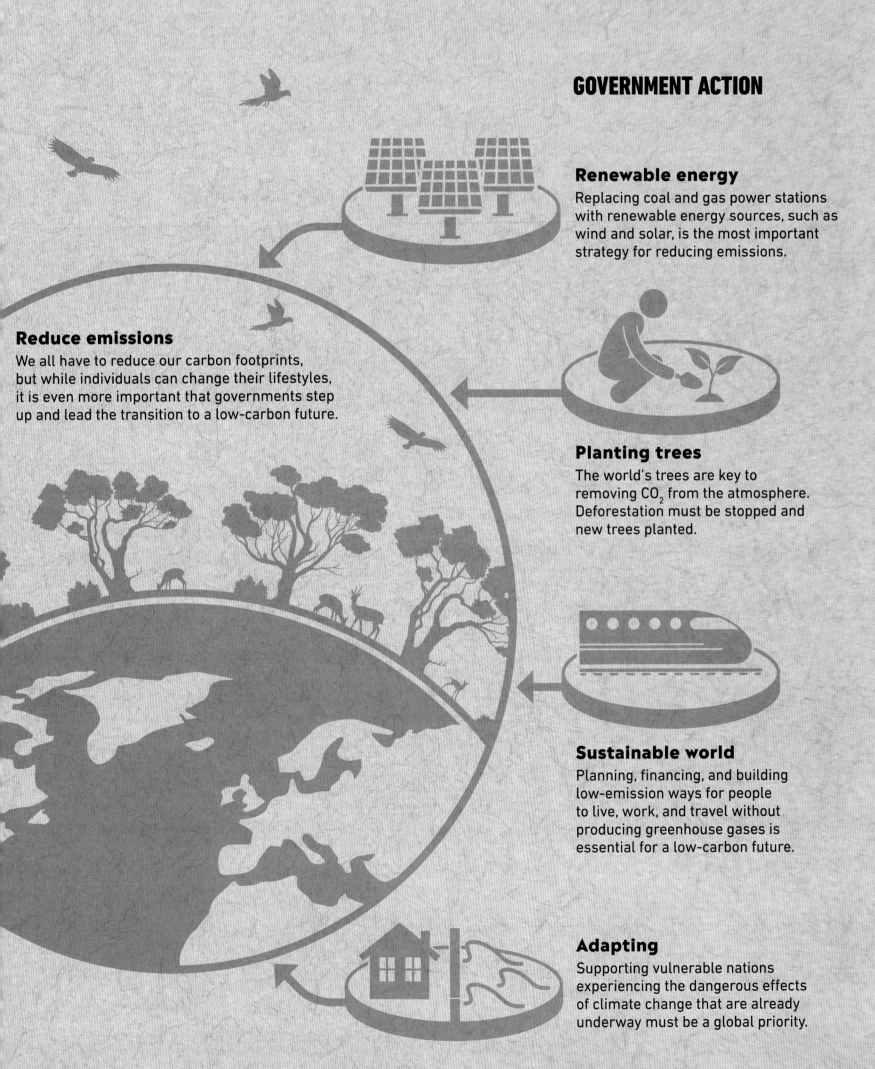

GOVERNMENT ACTION

Renewable energy
Replacing coal and gas power stations with renewable energy sources, such as wind and solar, is the most important strategy for reducing emissions.

Reduce emissions
We all have to reduce our carbon footprints, but while individuals can change their lifestyles, it is even more important that governments step up and lead the transition to a low-carbon future.

Planting trees
The world's trees are key to removing CO_2 from the atmosphere. Deforestation must be stopped and new trees planted.

Sustainable world
Planning, financing, and building low-emission ways for people to live, work, and travel without producing greenhouse gases is essential for a low-carbon future.

Adapting
Supporting vulnerable nations experiencing the dangerous effects of climate change that are already underway must be a global priority.

United States
The USA only uses a small share of renewables and contributes a large part of global greenhouse gas emissions. In 2017, it made plans to withdraw from the Paris Agreement, jeopardizing the initiative's success.

Iceland
Nearly all of the energy in this island nation already comes from hydropower and geothermal energy, but as Iceland is part of the EU, it needs to further limit emissions from its transport sector to keep its target.

The EU
The countries of the EU made a joint pledge to reduce emissions, but they have to do much more to reach their target.

Morocco
Morocco's high star rating comes from a combination of low emissions per capita, and its pledge to increase renewables to 52 per cent by 2050. The country needs to do more to phase out coal, but it is still on target for a rise of just 1.5°C.

Gambia
The only other country on track for 1.5°C, Gambia is investing in solar power and reforestation.

Ethiopia
Energy consumption is not very high in Ethiopia, and over 75 per cent of the country's energy comes from renewables, mostly hydroelectric and solar power. It is one of the few countries on track to meet the 2°C goal.

Brazil
Although it makes good use of renewable energy sources, such as hydropower, Brazil's rapid rate of deforestation and high methane emissions due to cattle farming mean it is not doing enough to keep its Paris Agreement pledge.

ACCORDING TO THE CLIMATE ACTION TRACKER ONLY TWO COUNTRIES ARE ON TRACK FOR A RISE OF 1.5°C

THE UK AIMS FOR NET-ZERO EMISSIONS BY 2050

Paris Agreement
In 2015, 197 countries met in Paris, France, to discuss climate change. They agreed that, at the least, we need to keep the global average temperature from rising by more than 2°C (3°F) above what it was before the Industrial Revolution – but aim for just a 1.5°C (2.7°F) rise, or ideally below that. To achieve this, all countries must minimize greenhouse gas emissions, and regularly review their targets.

Nations Unies
Conférence sur les Changements Climatiques 2015
COP21/CMP11
Paris, France

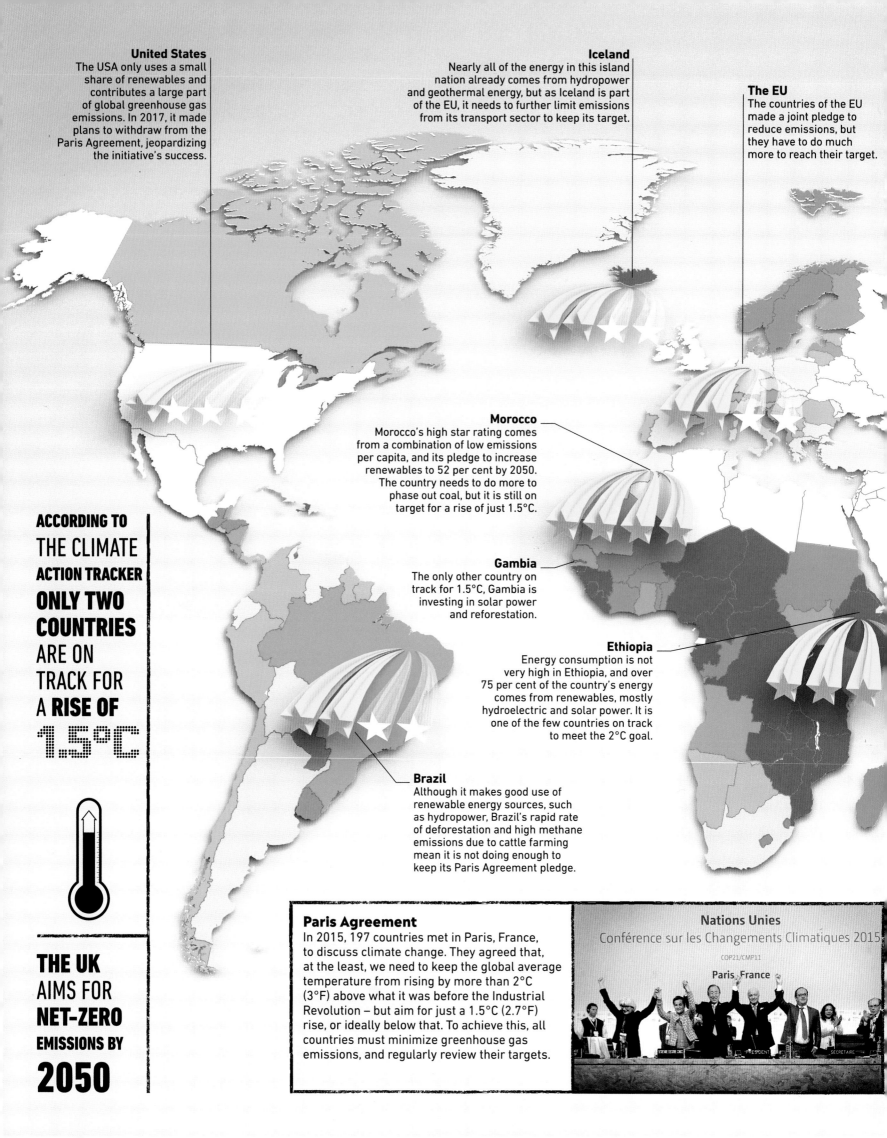

International action

All countries are responsible for setting their own targets to reduce emissions, but to achieve global progress they need to work together. International negotiations, such as the Paris Agreement, establish joint responsibility and a framework in which richer countries assist poorer countries financially so they can meet their targets.

KEY

The darker the green, the higher percentage of a country's final energy is produced from renewable sources. The star ratings shown for a selection of countries tells how much action they have taken to reach the Paris Agreement target.

- Below 10 per cent
- 10–20 per cent
- 20–30 per cent
- 30–50 per cent
- 50–70 per cent
- Above 75 per cent
- No data

★★★★ On track for 1.5°C target

★★★☆ On track for 2°C target

★★☆☆ Not doing enough

★☆☆☆ Critically insufficient

Russia
Russia has not invested as much in renewable energy sources as other countries. A massive effort is needed for it to get on track to meet its Paris Agreement targets.

China
Despite an enormous number of wind turbines and solar panels, China's emissions are extremely high due to its heavy use of coal. Polluting industries and a steep increase in fossil-fuelled car ownership contribute to its low star rating.

India
A newly emerging industrial nation with a huge population, India is making good use of renewables. It needs to continue to reduce its reliance on coal, and ensure there are enough charging stations to match its ambitious electric vehicle scheme, but it is on track to meet the 2°C target.

Australia
While lots of Australians are installing solar panels on their own homes, fossil fuels emissions from industry continue to rise and Australia is currently not meeting its target.

4.8°C (9.6°F) IS THE EXPECTED TEMPERATURE RISE IF WE DO NOTHING

Keeping it green

The green shades on this map show how much of each country's energy comes from sources like wind and solar power rather than fossil fuels. The star rating signals whether the effort being made by a country is enough to bring emissions in line with internationally agreed goals. This depends on many different factors, carefully analyzed and compared with the Paris Agreement targets by independent scientists for The Climate Action Tracker (CAT) review.

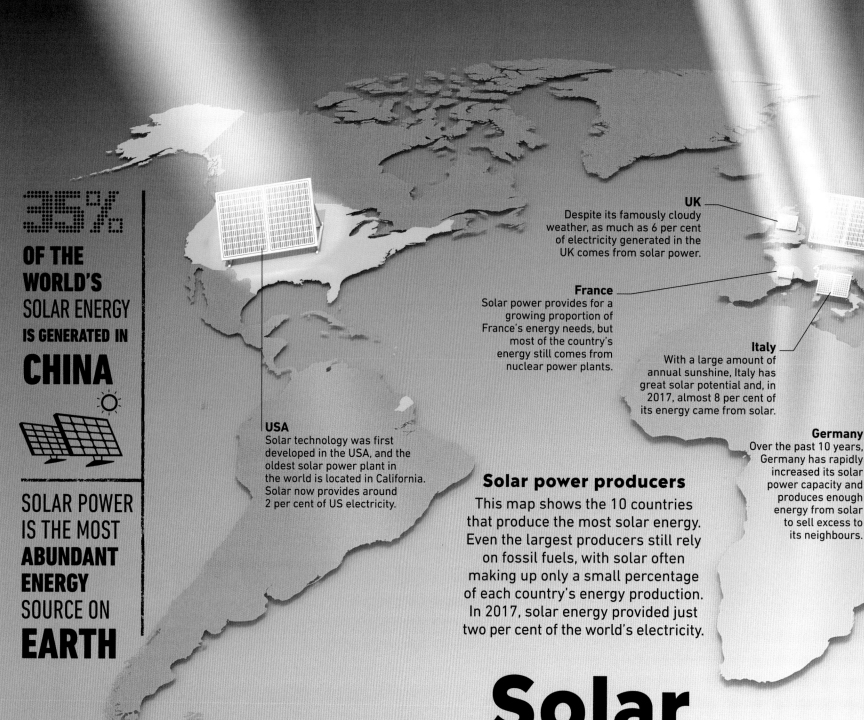

35%
OF THE WORLD'S SOLAR ENERGY IS GENERATED IN CHINA

SOLAR POWER IS THE MOST **ABUNDANT ENERGY** SOURCE ON **EARTH**

UK
Despite its famously cloudy weather, as much as 6 per cent of electricity generated in the UK comes from solar power.

France
Solar power provides for a growing proportion of France's energy needs, but most of the country's energy still comes from nuclear power plants.

Italy
With a large amount of annual sunshine, Italy has great solar potential and, in 2017, almost 8 per cent of its energy came from solar.

Germany
Over the past 10 years, Germany has rapidly increased its solar power capacity and produces enough energy from solar to sell excess to its neighbours.

USA
Solar technology was first developed in the USA, and the oldest solar power plant in the world is located in California. Solar now provides around 2 per cent of US electricity.

Solar power producers

This map shows the 10 countries that produce the most solar energy. Even the largest producers still rely on fossil fuels, with solar often making up only a small percentage of each country's energy production. In 2017, solar energy provided just two per cent of the world's electricity.

Solar solutions

Patriotic suntrap
China is the current world leader in solar energy and is continuing to invest in building new solar generating plants. In fact, more than 60 per cent of the world's solar panels are made in China, several of which were used to create this impressive solar park in the shape of a panda.

Energy captured from the Sun's rays offers a renewable source of power that never runs out. Even better, solar plants produce no carbon dioxide, so do not contribute to climate change. There are solar plants in countries across the world, but many more are needed to reduce our dependence on oil and gas for energy.

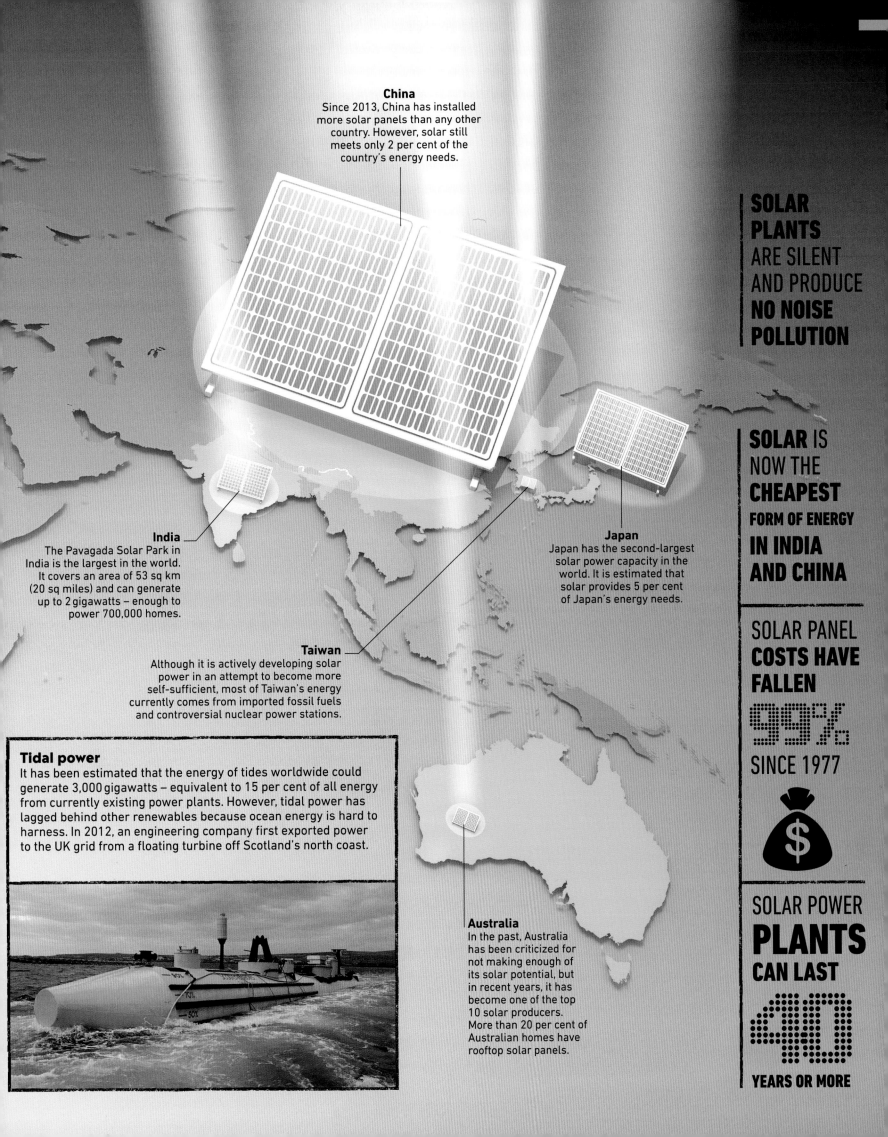

China
Since 2013, China has installed more solar panels than any other country. However, solar still meets only 2 per cent of the country's energy needs.

India
The Pavagada Solar Park in India is the largest in the world. It covers an area of 53 sq km (20 sq miles) and can generate up to 2 gigawatts – enough to power 700,000 homes.

Japan
Japan has the second-largest solar power capacity in the world. It is estimated that solar provides 5 per cent of Japan's energy needs.

Taiwan
Although it is actively developing solar power in an attempt to become more self-sufficient, most of Taiwan's energy currently comes from imported fossil fuels and controversial nuclear power stations.

Tidal power
It has been estimated that the energy of tides worldwide could generate 3,000 gigawatts – equivalent to 15 per cent of all energy from currently existing power plants. However, tidal power has lagged behind other renewables because ocean energy is hard to harness. In 2012, an engineering company first exported power to the UK grid from a floating turbine off Scotland's north coast.

Australia
In the past, Australia has been criticized for not making enough of its solar potential, but in recent years, it has become one of the top 10 solar producers. More than 20 per cent of Australian homes have rooftop solar panels.

SOLAR PLANTS ARE SILENT AND PRODUCE **NO NOISE POLLUTION**

SOLAR IS NOW THE **CHEAPEST** FORM OF ENERGY **IN INDIA AND CHINA**

SOLAR PANEL **COSTS HAVE FALLEN** 99% SINCE 1977

SOLAR POWER **PLANTS CAN LAST** 40 **YEARS OR MORE**

The Ivanpah solar power plant supplies electricity to 140,000 homes in Southern California.

Harnessing solar power

Solar energy fulfils only a small proportion of the world's energy needs, but this is set to change as countries invest in new technologies. Opened in 2013, the Ivanpah solar power plant sprawls across 14 sq km (5.5 sq miles) of the USA's Mojave Desert. It is the world's largest concentrated solar power facility, using mirrors rather than solar panels. Ivanpah's more than 300,000 mirrors are arranged around three 140-m- (450-ft-) high towers. The mirrors are controlled by computer to track the Sun and reflect the light towards a huge boiler at the top of each tower. The concentrated sunlight heats water in the boiler, turning the water to steam, which drives an electricity-generating turbine at the tower's base.

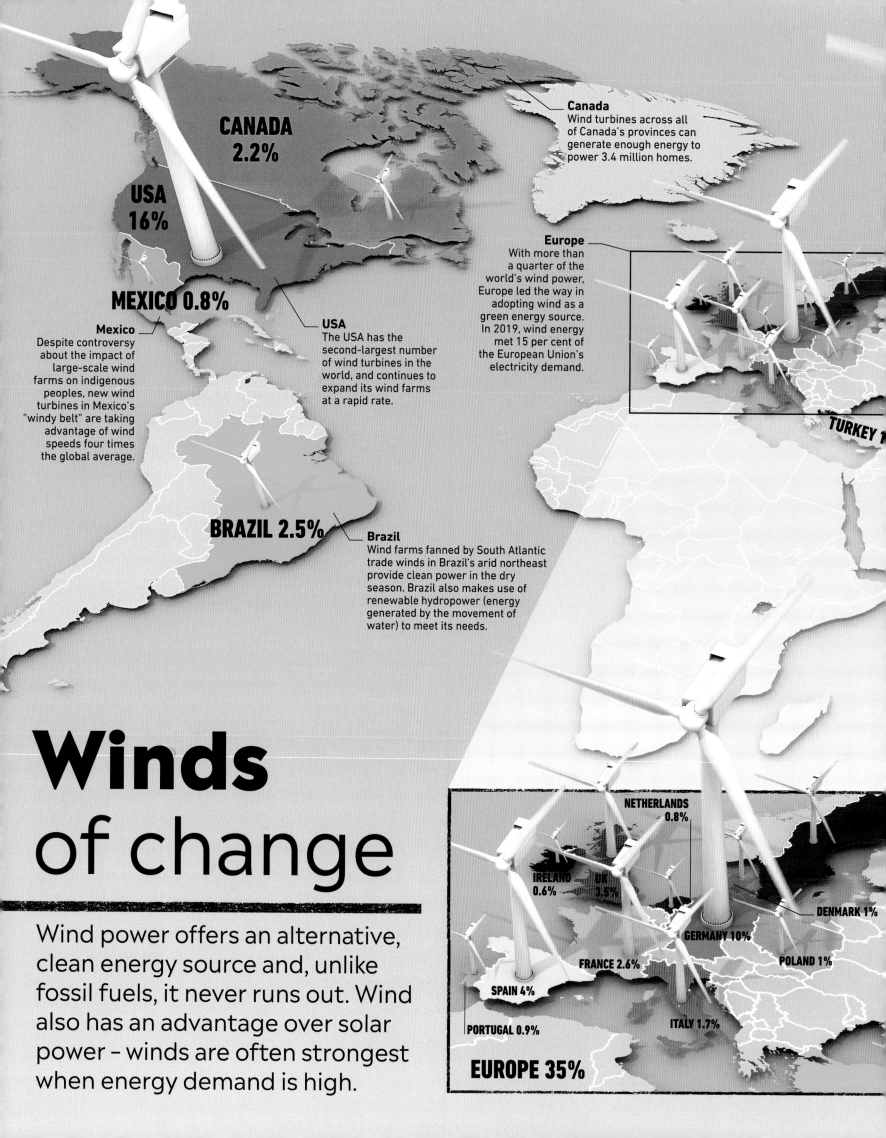

CANADA 2.2%

Canada
Wind turbines across all of Canada's provinces can generate enough energy to power 3.4 million homes.

USA 16%

MEXICO 0.8%

Mexico
Despite controversy about the impact of large-scale wind farms on indigenous peoples, new wind turbines in Mexico's "windy belt" are taking advantage of wind speeds four times the global average.

USA
The USA has the second-largest number of wind turbines in the world, and continues to expand its wind farms at a rapid rate.

Europe
With more than a quarter of the world's wind power, Europe led the way in adopting wind as a green energy source. In 2019, wind energy met 15 per cent of the European Union's electricity demand.

TURKEY 1

BRAZIL 2.5%

Brazil
Wind farms fanned by South Atlantic trade winds in Brazil's arid northeast provide clean power in the dry season. Brazil also makes use of renewable hydropower (energy generated by the movement of water) to meet its needs.

Winds
of change

Wind power offers an alternative, clean energy source and, unlike fossil fuels, it never runs out. Wind also has an advantage over solar power – winds are often strongest when energy demand is high.

NETHERLANDS 0.8%

IRELAND 0.6%

UK 3.5%

DENMARK 1%

GERMANY 10%

FRANCE 2.6%

POLAND 1%

SPAIN 4%

ITALY 1.7%

PORTUGAL 0.9%

EUROPE 35%

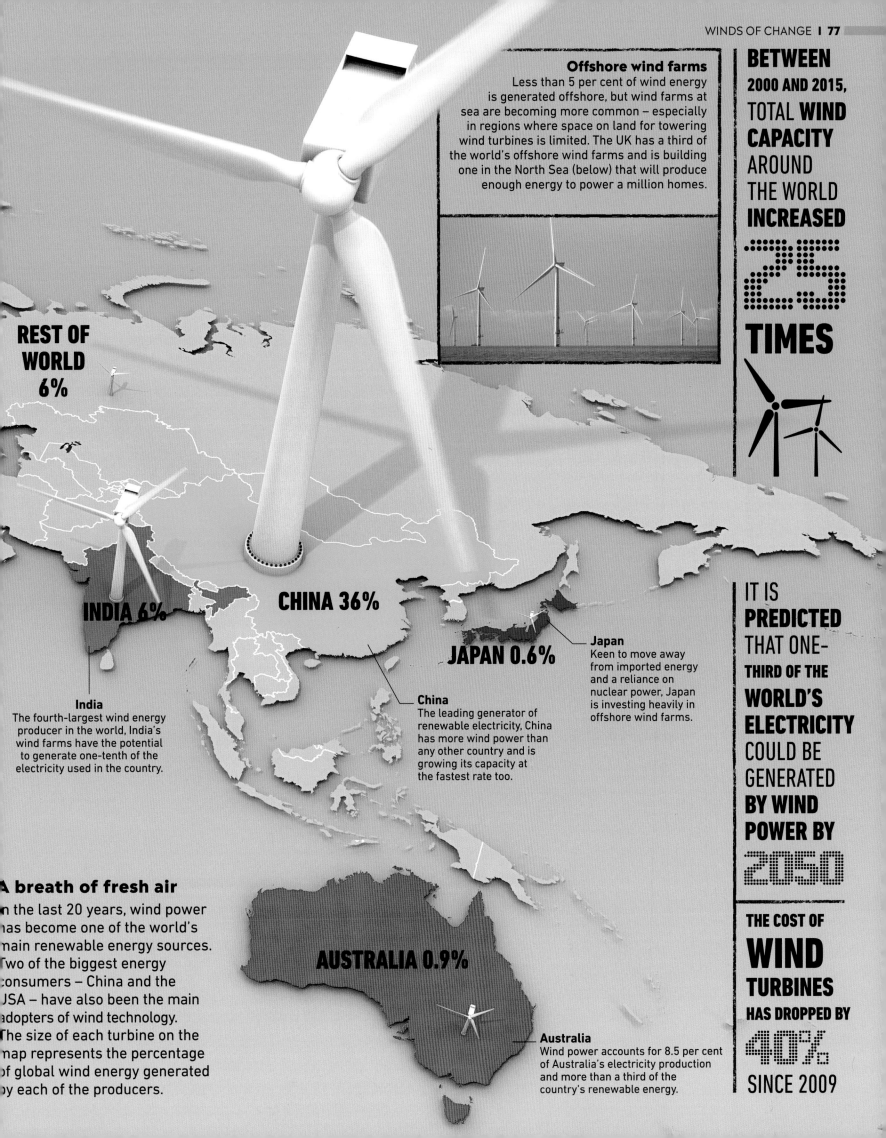

Offshore wind farms
Less than 5 per cent of wind energy is generated offshore, but wind farms at sea are becoming more common – especially in regions where space on land for towering wind turbines is limited. The UK has a third of the world's offshore wind farms and is building one in the North Sea (below) that will produce enough energy to power a million homes.

BETWEEN 2000 AND 2015, TOTAL **WIND CAPACITY** AROUND THE WORLD INCREASED **25 TIMES**

REST OF WORLD 6%

INDIA 6%

CHINA 36%

JAPAN 0.6%

Japan
Keen to move away from imported energy and a reliance on nuclear power, Japan is investing heavily in offshore wind farms.

India
The fourth-largest wind energy producer in the world, India's wind farms have the potential to generate one-tenth of the electricity used in the country.

China
The leading generator of renewable electricity, China has more wind power than any other country and is growing its capacity at the fastest rate too.

IT IS **PREDICTED** THAT ONE-THIRD OF THE **WORLD'S ELECTRICITY** COULD BE GENERATED **BY WIND POWER BY 2050**

A breath of fresh air
In the last 20 years, wind power has become one of the world's main renewable energy sources. Two of the biggest energy consumers – China and the USA – have also been the main adopters of wind technology. The size of each turbine on the map represents the percentage of global wind energy generated by each of the producers.

AUSTRALIA 0.9%

Australia
Wind power accounts for 8.5 per cent of Australia's electricity production and more than a third of the country's renewable energy.

THE COST OF **WIND TURBINES** HAS DROPPED BY **40%** SINCE 2009

1. Planting more than promised, USA
The United States is among the world's highest emitters of CO_2. It made a Bonn Challenge pledge to plant 150,000 sq km (58,000 sq miles) of trees by 2020 and has already planted more than its target.

2. Rainforest restoration, Brazil
In addition to the threatened Amazon, Brazil's Atlantic rainforest is also in urgent need of help. Ambitious projects are taking place but progress has been slowed due to backtracking on political promises.

3. Underwater meadow, UK
Meadows of seagrass, plants that grow underwater, form important carbon sinks. Projects like this one off the coast of Wales, UK, aim to reseed and protect seagrass meadows damaged by human impact.

4. Tree nursery, Sweden
Some of the countries that depend on their forestry industry make sure they plant more trees than they fell. In the last 100 years, Sweden has replaced its lost trees, and now almost 70 per cent of the country is forested.

 **350** MILLION TREES WERE PLANTED IN 12 HOURS AS PART OF ETHIOPIA'S **FOUR-BILLION TREES PLEDGE**

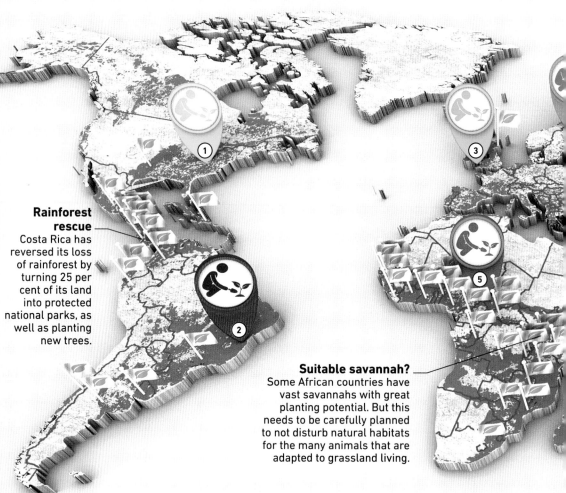

Rainforest rescue
Costa Rica has reversed its loss of rainforest by turning 25 per cent of its land into protected national parks, as well as planting new trees.

Suitable savannah?
Some African countries have vast savannahs with great planting potential. But this needs to be carefully planned to not disturb natural habitats for the many animals that are adapted to grassland living.

Restoring forests

Around the world, organizations and volunteers are planting new trees. Forests naturally capture carbon dioxide (CO_2) from the air, which makes them a carbon sink. Many of the projects highlighted here also help communities and ecosystems adapt to the consequences of climate change.

THE COUNTRIES WITH THE BEST REFORESTATION POTENTIAL ARE RUSSIA, THE UNITED STATES, CANADA, AUSTRALIA, BRAZIL, AND CHINA

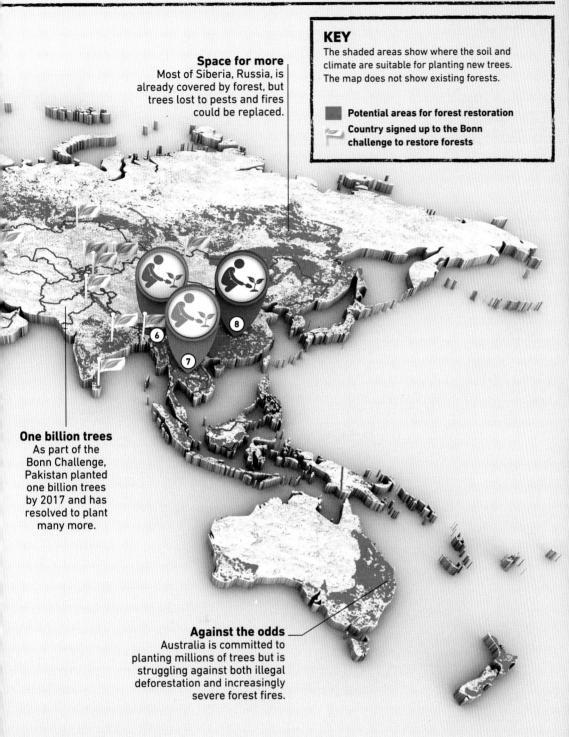

Space for more
Most of Siberia, Russia, is already covered by forest, but trees lost to pests and fires could be replaced.

KEY
The shaded areas show where the soil and climate are suitable for planting new trees. The map does not show existing forests.

Potential areas for forest restoration

Country signed up to the Bonn challenge to restore forests

One billion trees
As part of the Bonn Challenge, Pakistan planted one billion trees by 2017 and has resolved to plant many more.

Against the odds
Australia is committed to planting millions of trees but is struggling against both illegal deforestation and increasingly severe forest fires.

Planting potential

This map shows where it should be possible to plant more trees, usually in areas where there is already some forest, or where forests recently stood. It also shows the 54 countries that have so far signed up for the Bonn Challenge, each marked by a flag. Begun in 2011, this initiative aims to restore 350 million hectares of forest – equivalent to 490 million football pitches – by 2030. Although reducing emissions is the most important means to reach "net zero", the point at which the amount of greenhouse gases emitted is balanced by the amount removed from the atmosphere, planting forests can speed up our race to get there.

5. Great Green Wall, Sahel region, Africa
More than 20 African countries along the southern edge of the Sahara desert, from Senegal to Ethiopia, are planting a wall of trees, nearly 8,000 km (5,000 miles) long, to capture carbon and stop the desert spreading.

6. Elephant corridor, Assam, India
Sometimes, trees are planted as part of creating wildlife corridors – safe passages to be used by wild animals moving between sections of forests without crossing roads or trampling over farmland.

7. Mangrove reforestation, Thailand
Mangrove forests are not just good carbon sinks, they also protect shores from flooding and erosion, and are home to many animals. Planting projects such as this one in Thailand take place in many tropical regions.

8. Grain for Green programme, China
Chinese farmers are paid to plant trees on vast areas of land once cleared to grow food. The programme has been successful, but often only one type of tree is planted, which is not good for animal and plant diversity.

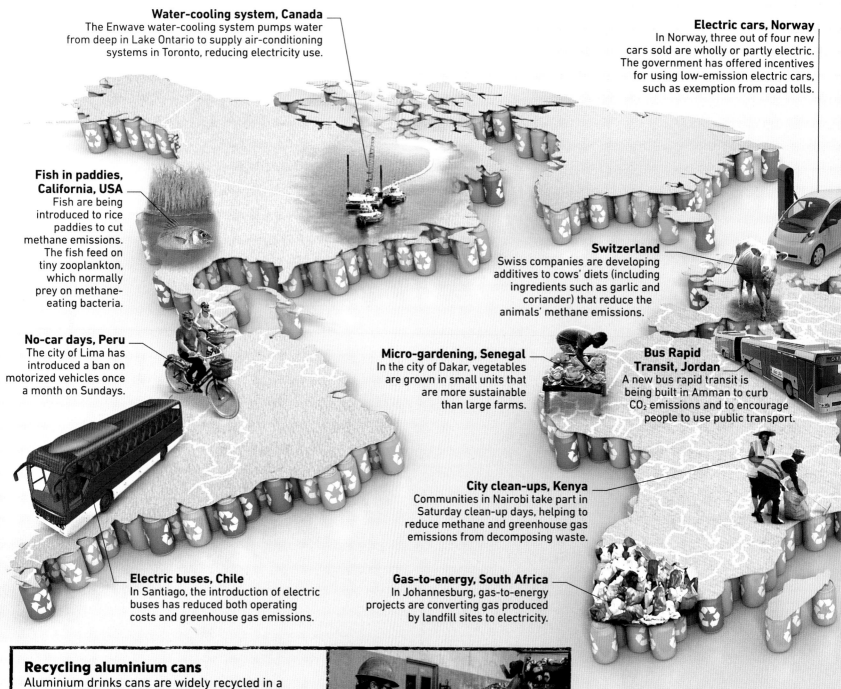

Water-cooling system, Canada
The Enwave water-cooling system pumps water from deep in Lake Ontario to supply air-conditioning systems in Toronto, reducing electricity use.

Electric cars, Norway
In Norway, three out of four new cars sold are wholly or partly electric. The government has offered incentives for using low-emission electric cars, such as exemption from road tolls.

Fish in paddies, California, USA
Fish are being introduced to rice paddies to cut methane emissions. The fish feed on tiny zooplankton, which normally prey on methane-eating bacteria.

Switzerland
Swiss companies are developing additives to cows' diets (including ingredients such as garlic and coriander) that reduce the animals' methane emissions.

No-car days, Peru
The city of Lima has introduced a ban on motorized vehicles once a month on Sundays.

Micro-gardening, Senegal
In the city of Dakar, vegetables are grown in small units that are more sustainable than large farms.

Bus Rapid Transit, Jordan
A new bus rapid transit is being built in Amman to curb CO_2 emissions and to encourage people to use public transport.

City clean-ups, Kenya
Communities in Nairobi take part in Saturday clean-up days, helping to reduce methane and greenhouse gas emissions from decomposing waste.

Electric buses, Chile
In Santiago, the introduction of electric buses has reduced both operating costs and greenhouse gas emissions.

Gas-to-energy, South Africa
In Johannesburg, gas-to-energy projects are converting gas produced by landfill sites to electricity.

Recycling aluminium cans
Aluminium drinks cans are widely recycled in a process that saves more greenhouse gas emissions than the recycling of any other household rubbish. This is because the melting of scrap aluminium to create new products uses about five per cent of the energy needed to make new aluminium.

Green future
This map shows some of the sustainability initiatives in progress worldwide that are helping to reduce greenhouse gas emissions.

Changing how we live

Around the world people are looking for innovative ways to reduce human impact on the climate. This idea, known as sustainability, is about adjusting our lives so that we become less reliant on practices that will spoil the world for future generations.

Circular fashion

The world fashion industry is beginning to consider ways to reduce waste and greenhouse gas emissions through a concept called circular fashion. Instead of encouraging the purchase of new items every season, clothes and shoes will be made from durable materials that have been produced in sustainable ways. Older garments will be repaired, recycled, swapped, rented, or sold second-hand, instead of going straight into landfill.

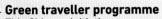

96

CITIES ARE PART OF THE C40 CITIES GROUP, WHICH IMPLEMENTS GREEN PROJECTS

Green traveller programme
This Chinese initiative encourages people to leave their cars at home by paying them a small amount for every day they don't drive.

FIFA World Cup, Qatar
The 2022 FIFA World Cup stadiums will be lit by energy-efficient LED lights. One stadium will be made from shipping containers that can be dismantled after the event.

LED lighting, India
The use of LED energy-efficient light bulbs has increased rapidly in India in recent years. Sales rose from 5 million bulbs per year in 2014 to 670 million in 2018.

Recycling fishing nets, Philippines
Communities are recycling discarded plastic fishing nets and turning them into carpets.

Bullet trains, Japan
The modern Shinkansen Bullet Train has a beak-shaped front to make it more streamlined and runs on 15 per cent less electricity than previous models.

Multiple-use plastics, Australia
The city of Darwin is phasing out items made from single-use plastics, such as coffee cups and straws, from events and markets on council land. People are encouraged to bring their own reusable cups instead.

Greener city, Australia
The city of Melbourne plans to create more green areas as well as build new carbon-neutral housing projects.

BETWEEN 200 AND 350 BILLION DRINKS CANS ARE USED EACH YEAR AND **70%** ARE RECYCLED

Electric trains, New Zealand
Major funding is going into a new electric train network in Auckland.

 GERMANY RECYCLES ABOUT **65%** OF ITS RUBBISH – **MORE THAN** ANY OTHER **NATION**

Planet-friendly eating

The farming of animals for meat and dairy foods is a major source of greenhouse gases (GHGs). One study has found that adopting a vegan diet, which cuts out all animal foods, could cut GHGs from food by half. Eating and wasting less food also lowers emissions.

Cheese
Cows produce a lot of methane, and so dairy milk has much higher emissions than any plant-based alternatives. It takes 10 l (20 pints) of milk to make 1 kg (2 lb) of cheese, which means it has a big environmental impact.

Eggs
Protein-rich and nutrient-dense eggs have relatively low emissions when compared to meat. Yet they still come at a high cost to the environment if compared to pulses, nuts, grains, and other plant-based foods.

What should I eat?

These four plates compare the amount of GHGs emitted in the production of the food found on them. The size of the dome indicates the amount of emissions. Each portion is enough to provide 50 g (1.8 oz) of protein, the recommended daily allowance.

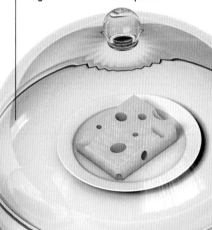

CHEESE 5.4 KG GHGs

EGGS 2.1 KG GHGs

Beans
Changing to a diet based on pulses, such as beans and peas, whole grains, nuts, seeds, vegetables, and fruit can massively reduce human impact on the climate. Emissions from plant-based foods are far lower than from animal-based foods.

BEANS 0.4 KG GHGs

Meatless meat
Today, there is a growth in new foods that offer an alternative to meat. Some look similar to meat products, such as burgers, chicken nuggets, and sausages, but are made from soya or mycoprotein, a naturally occurring fungus. Use of the term "meat" for these highly processed products is controversial, but their taste and texture are designed to appeal to meat-eaters and may promote sustainable eating habits.

North America
The average American consumes a whopping 124 kg (273 lb) of meat a year. North America produces twice as much meat as 60 years ago.

South America
Meat consumption has grown most in countries with the fastest economic growth, such as Brazil, where people eat an average of 100 kg (220 lb) of meat a year.

  6% OF **GREENHOUSE GAS EMISSIONS GLOBALLY** ARE FROM **FOOD WASTE**

Beef
Meat has the highest emissions of any food, and beef more than any other. Poultry, pork, and even lamb have lower emissions, so switching from steak to chicken is one way to reduce your carbon footprint.

BEEF 25 KG GHGs

Reducing food waste
Around one-third of food produced globally goes to waste – enough to feed the one billion people in the world who go hungry. Left to rot on farms, spoiled on its way to market, or binned in shops, cafés, and homes, this wasted food generates greenhouse gases in production and, if it ends up in landfill, as it decomposes. A less wasteful approach to food would help feed the world's growing population without increasing GHG emissions.

KEY
GHG emissions from meat consumption per capita.

- below 500 kg (½ ton)
- 500–1000 kg (½–1 ton)
- 1000–1500 kg (1–1½ tons)
- 1500–2000 kg (1½–2 tons)
- 2000–2500 kg (2–2½ tons)
- above 2500 kg (2½ tons)
- No data

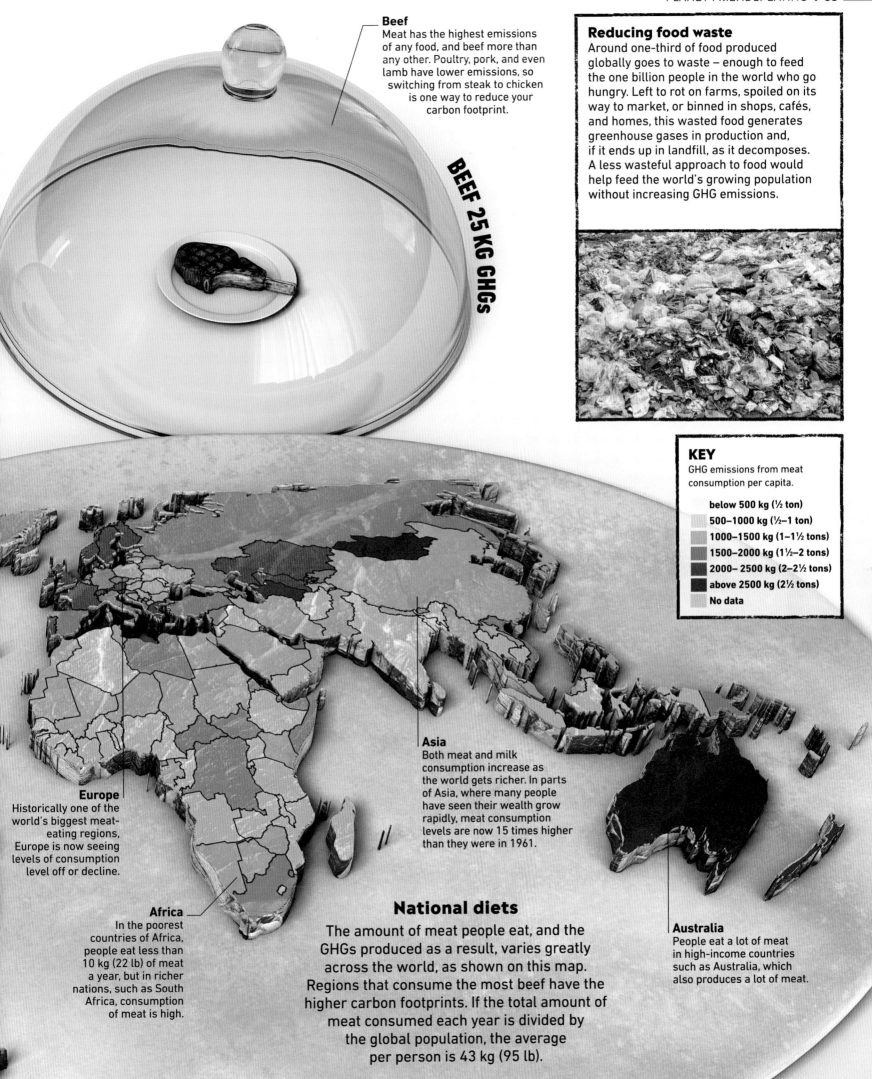

Asia
Both meat and milk consumption increase as the world gets richer. In parts of Asia, where many people have seen their wealth grow rapidly, meat consumption levels are now 15 times higher than they were in 1961.

Europe
Historically one of the world's biggest meat-eating regions, Europe is now seeing levels of consumption level off or decline.

Africa
In the poorest countries of Africa, people eat less than 10 kg (22 lb) of meat a year, but in richer nations, such as South Africa, consumption of meat is high.

National diets
The amount of meat people eat, and the GHGs produced as a result, varies greatly across the world, as shown on this map. Regions that consume the most beef have the higher carbon footprints. If the total amount of meat consumed each year is divided by the global population, the average per person is 43 kg (95 lb).

Australia
People eat a lot of meat in high-income countries such as Australia, which also produces a lot of meat.

Flood defences in New Orleans, USA

Low-lying New Orleans has adapted to sea-level rise and more frequent and intense storms by building a massive system of levees, flood walls, and storm surge barriers around the city. Built with $14.5 billion of US government funding, the works reduce hurricane risk but may not protect residents from future increases in water levels.

Storm surge barriers in the Netherlands

Around two-thirds of the Netherlands, an area home to nine million people, is at risk from rising sea levels. The Eastern Scheldt storm surge barrier is one of 13 moveable flood defences that the country has built along its coast and river deltas for protection, with the aim of preserving the life of every individual and the economy.

Adapting to climate change

Around the world, countries are having to respond to climate change, but the ability to adapt is unequal. The rich countries that have emitted most GHGs can afford large-scale climate-adaptation strategies, while poorer countries have to find solutions with limited resources.

A fairer future?

As countries become wealthier, they produce more greenhouse gases (GHGs). Richer countries are also able to spend more money to protect themselves against the effects of climate change. To tackle the climate emergency, richer countries must support the poorer, who are often the most likely to experience the worst effects of the climate crisis and have done the least to cause it.

California, USA
Even in the wealthiest of nations, simple adaptations can be best. In California, grazing goats are creating fire breaks as a protection against wildfires.

NORTH AMERICA 30%

Costa Rica
In Costa Rica, farmers are switching from growing coffee to oranges as the country experiences drought, which make its climate less suited to coffee plantations.

SOUTH AMERICA 3%

Indigenous management of palm swamps, Peru

Native palm swamps in Peru's Datem del Marañón province are affected by flooding and drought as a result of deforestation in the Amazon basin. Indigenous communities are managing the wetlands in ways that enable local people to live sustainably without felling trees. This protects a natural carbon sink and preserves biodiversity.

Fresh water in Dakar, Senegal

Coastal Senegal is vulnerable to rising sea levels, drought, and torrential rains. The poorest parts of its capital, Dakar, are most at risk from floods, while encroaching saltwater threatens freshwater supplies. Improving drainage, building dykes and ponds to manage water, and using salt-adapted seeds are ways to tackle these hazards.

Green roof gardens in Shanghai, China
China's megacities have sprung up very fast and often with limited urban planning. Heatwaves can make them unbearable as the Sun's rays reflect off glass and concrete. To reduce this urban heat island effect, city planners are making policies to increase green spaces and build roof gardens, such as these "hanging gardens" in Shanghai.

Wineries in the Murray Valley, Australia
Drought has affected grape harvests in southeast Australia for several years. Some wineries now harvest grapes earlier, while others have either abandoned vines or relocated to less dry areas. Communities along the Murray Darling River are making commitments to save and share water as it becomes a scarce resource.

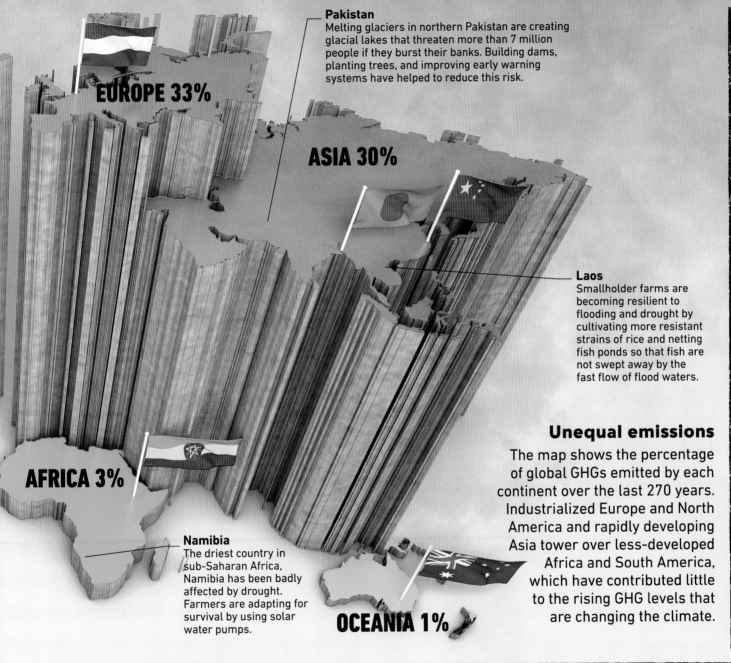

Pakistan
Melting glaciers in northern Pakistan are creating glacial lakes that threaten more than 7 million people if they burst their banks. Building dams, planting trees, and improving early warning systems have helped to reduce this risk.

EUROPE 33%

ASIA 30%

Laos
Smallholder farms are becoming resilient to flooding and drought by cultivating more resistant strains of rice and netting fish ponds so that fish are not swept away by the fast flow of flood waters.

AFRICA 3%

Namibia
The driest country in sub-Saharan Africa, Namibia has been badly affected by drought. Farmers are adapting for survival by using solar water pumps.

OCEANIA 1%

Unequal emissions

The map shows the percentage of global GHGs emitted by each continent over the last 270 years. Industrialized Europe and North America and rapidly developing Asia tower over less-developed Africa and South America, which have contributed little to the rising GHG levels that are changing the climate.

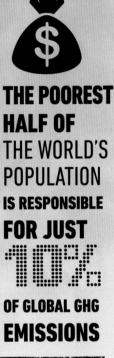

THE POOREST HALF OF THE WORLD'S **POPULATION IS RESPONSIBLE FOR JUST** 10% **OF GLOBAL GHG EMISSIONS**

80% **OF PEOPLE IN THE WORLD HAVE NEVER FLOWN**

Drought-resistant grass in Ethiopia's highlands
African countries have contributed the least to climate change but are expected to be among the hardest hit. Drought is a risk for Ethiopia's arid highlands, which are particularly exposed to a lack of rain. Local micro-businesses are supplying drought-tolerant grass seeds to farm communities to grow animal feed and restore degraded land.

Cyclone shelters in Bangladesh
Around 18 million Bangladeshis live in areas vulnerable to cyclones and sea-level rise. Shelters and early warning systems help people weather storms but coastal defences, such as shoring up riverbanks with sandbags, are still low-tech. People are trying to protect their homes however they can, but some are forced to seek work in the cities.

The city of Copenhagen has reduced its CO_2 emissions by 40% since 2005.

Fume-free energy

Amager Resource Centre, also known as Copenhill, in Copenhagen, Denmark, is a waste incinerator with a difference. Treating up to 400,000 tonnes of waste a year, it burns rubbish to generate electricity for 50,000 households and provide heat for up to 120,000 more. The smoke created is cleaned of harmful particles so that the chimney stack releases clouds of harmless water vapour. One of the cleanest plants of its kind in the world, it doubles as a mountain sports centre, with a 400-m (1,300-ft) ski slope on the roof. Reducing emissions and producing sustainable energy is a central part of Copenhagen's mission to become the first carbon-neutral capital city by 2025.

Worldwide demonstrations

This map shows some of the climate protests that have taken place across the world, including many local protests.

Energy East Pipeline, Canada
Plans for a pipeline to transport 1.1 million barrels of oil a day were abandoned in 2017, partly in response to widespread protests.

Climate Strike, Sweden
Thousands of people across Sweden protested during the Global Week for Future in September 2019.

Dakota Access Pipeline, USA
Climate campaigners protested against the building of a pipeline carrying 470,000 barrels of oil a day. Despite this, the pipeline was completed in 2017.

Amazon protest, Brazil
In August 2019, protests took place across Brazil against fires in the Amazon, which were caused in part by deforestation.

Heathrow expansion, UK
For years, environmental activists have demonstrated against plans to build a third runway at Heathrow Airport.

Lamu Coal Plant, Kenya
In 2019, protestors in Kenya stopped the development of the nations' first coal power plant, proposed for the Lamu region.

Guaiba, Brazil
In February 2020, protestors successfully opposed development of the largest open-pit coal mine in South America.

Greta Thunberg
Swedish environmental campaigner Greta Thunberg has encouraged millions to become involved in the climate protest movement. In 2018, aged just 15, she skipped school to protest against climate change outside the Swedish parliament building. Since then, she has campaigned on the world stage for governments to take immediate steps to address the climate crisis.

Climate Strike, Argentina
In September 2019, crowds marched through Buenos Aires demanding urgent action on the global climate crisis.

"Coal Kills", South Africa
Activists protesting against fossil fuels gathered outside the Southern African Coal Conference in Cape Town, in January 2020.

 STUDENT STRIKES HAVE TAKEN PLACE IN SOME **228 COUNTRIES** ACROSS THE WORLD

  MILLION PEOPLE TOOK PART IN THE **SEPTEMBER 2019** CLIMATE STRIKES WORLDWIDE

Planet protest

In recent years, a mass global movement has evolved, calling for action on climate change. Millions of people are communicating on social media and protesting on the streets.

Single protests, Russia
In Moscow in 2019, activists queued each Friday to stage single-person protests, as group demonstrations are restricted by the Russian government.

Forest fires, Russia
Demonstrators in the Siberian city of Krasnoyarsk demanded decisive action to tackle forest fires in 2019.

No Coal Japan
The No Coal Japan campaign is against government and business plans to expand coal power plants. The group staged a protest in June 2019, ahead of the G20 Summit in Osaka.

Plastic protest, India
Throughout 2019 there were demonstrations in India against single-use plastics, whose production releases greenhouse gases.

Red Rebels, Australia
The Red Rebel Brigade, dressed in long red robes (symbolizing blood), staged a protest at the Sydney Opera House in December 2019. Many protestors came from areas affected by wildfires.

Fridays for Future
Inspired by Greta Thunberg's protests, Fridays for Future (FFF) is a global movement in which school students take time off school on Fridays to participate in demonstrations (pictured, right, in Canada). These have often been timed to influence major summits, such as the United Nations Climate Action Summit in September 2019.

FRIDAYS FOR FUTURE WANTS THE WORLD TO ADHERE TO THE **GLOBAL TEMPERATURE TARGETS** SET OUT IN THE **2015 PARIS AGREEMENT**

What **you** can do

There are steps all of us can take to help to reduce the build up of greenhouse gases (GHGs) in the atmosphere. These range from small acts like turning off lights, to bigger lifestyle changes, or working to raise awareness. These changes might not feel like much, but together the actions of millions of people can have an effect.

Saving energy

There are lots of ways to save energy around the home, and when out and about. Many also save you money, too!

- Use energy-efficient light bulbs and appliances, unplug anything that drains electricity when not in use, and turn out the lights when you leave a room.
- Add layers in the winter instead of turning up the heat, and try to use fans instead of air conditioning in the summer.
- Consider how much water you use and try to use it efficiently.
- Walk or ride a bike to get around your local area if you can.
- For longer journeys, use public transport instead of getting in the car and driving.

Growing green

You can "live green" by making conscious choices that help cut down the amount of GHGs that enter the atmosphere.

- If you can, plant trees in your garden or as part of a scheme in your neighbourhood. If you can't, support projects planting trees around the world.
- Buy locally grown produce wherever possible – imported foods will have a larger carbon footprint.
- Buy paper products that are grown or gathered from sustainably managed forests.
- If you can, grow your own fruits and vegetables.
- Try to only eat produce that is in season – this cuts down on the transportation of out-of-season produce from around the world.
- Eat more vegetables and plant-based foods and less meat.
- Get a compost bin – this cuts down on methane emissions from rotting food waste.

Reduce, reuse, recycle

GHGs are emitted when sourcing raw materials for products, as well as manufacturing, transporting, and disposing of them.

- Choose things that last, so you buy fewer of them.
- Instead of buying new clothes, shop in vintage and second-hand stores.
- Carry reusable cups and water bottles with you.
- Choose products and packaging that can be recycled wherever possible.
- If an item is broken, don't throw it away – try to fix it.

Staying informed

It's important to understand the science and issues around climate change, be informed about ideas and initiatives, and find out your own impact on the planet.

- **Work out your carbon footprint, and that of your family, with an online carbon calculator. You can find one here: footprint.wwf.org.uk.**
- **If there are emissions you cannot reduce, consider offsetting (compensating for) them. One way this can be done is by paying carbon-offset organizations that support renewable energy or tree-planting projects.**
- **Study science in school to help you understand the ideas behind climate change. This will help you explain the issues to others and make the right decisions for yourself.**
- **Follow current affairs – news reports will keep you informed about how climate change is affecting life across the world, and the latest action that is being taken to combat the effects.**

Sharing your worries

Thinking about the climate emergency can lead to feelings of stress and anxiety.

- **Never worry on your own – talk about your fears and feelings with friends, family, or teachers.**
- **Remember that no one can solve the climate crisis on their own, and other people and organizations can be a great support.**
- **It can help if you are able to find something positive to do. There are lots of successful initiatives out there to inspire you. Even small actions can make a difference.**

Getting involved

To make a meaningful impact on climate change governments and big industries must take action, too. There are many ways you can get involved to demand change at this higher level.

- **Write to local and national politicians to let them know that climate change policies are a priority for you.**
- **Find your local climate change group, or start one in your area. You may find this website helpful: globalclimatestrike.net.**
- **Join or set up a climate change action group in your school.**

Glossary

Agriculture
The use of land to grow crops or raise domestic animals for food.

Arable land
Land on which it is possible to grow crops.

Atmosphere
A layer of gases surrounding Earth.

Biodegradable
Description of something that decays naturally, avoiding becoming a pollutant.

Biomass
Material made from living things such as plants or animals, whose energy can be used to make biofuel.

Carbon
A chemical element which occurs in gaseous carbon dioxide, as well as fossil fuels and biomass.

Carbon cycle
The flow of carbon between the atmosphere, ocean and land.

Carbon dioxide (CO$_2$)
A gas that is formed when carbon combines with oxygen. Carbon dioxide is absorbed by plants and emitted by burning fossil fuels. It is a greenhouse gas, and the largest contributor to global warming.

Carbon footprint
The amount of carbon dioxide (or equivalent) emissions for which an individual or activity is responsible.

Carbon neutral
A process or activity that results in a net zero release of carbon dioxide into the atmosphere, including using carbon offsetting.

Carbon offsetting
Compensating for carbon dioxide emissions into the atmosphere by supporting or taking part in schemes dedicated to making equivalent reductions, such as planting trees.

Carbon sink
A natural environment that absorbs and traps carbon, reducing the concentration of carbon dioxide in the atmosphere. Earth's two main carbon sinks are forests and oceans.

Clean energy
See renewable energy

Climate
The average weather conditions of an area over time (30 years).

Climate change
A change in the state of the climate, that persists for an extended period. In this book, climate change refers to changes since the mid 20th century, which have been attributed to human activity, as defined by the UNFCCC.

Deforestation
The cutting down of forests for timber or to clear land.

Development
The economic and social process by which societies become more wealthy. As a country develops, the average income of its citizens usually increases.

Ecosystem
A community of organisms in a specific environment that interact with and affect one another.

Emissions
The discharge of tiny particles, liquid vapour, or gases into the atmosphere.

Energy
What makes things happen. Energy can't be created or destroyed – it can only be converted to another form. For example, the chemical energy from burning fossil fuels can be converted into electrical energy, which is used to power lights, appliances, and other devices.

Enteric fermentation
The process by which ruminant animals (animals that chew cud regurgitated from the stomach) such as cattle and sheep digest food, producing methane as a byproduct.

Environment
Area in which plants, animals, and people live.

Fertilizer
A natural or chemical substance used to provide plants with additional nutrients.

Fossil fuels
Coal, oil, and gas, which are made from decomposed organisms that lived millions of years ago.

Geothermal

Heat stored beneath Earth's surface.

Glacier

A slow-moving mass of ice formed by the build-up of snow over a long period of time.

Global warming

The rise in average temperatures throughout the world. Global warming since the Industrial Revolution has been attributed to human activity altering the composition of the atmosphere.

Green energy

See renewable energy

Greenhouse gas

A gas that traps heat within the atmosphere. Greenhouse gases include carbon dioxide, methane, and nitrous oxide.

Hydropower

Energy from the movement of water that is converted into electricity.

Industrial Revolution

A period of time beginning in the 18th century when machines began to appear in factories and were used to make goods.

Landfill

Disposing of waste by burying it or filling in excavated land. Rotting food and other organic waste in landfills account for a large proportion of methane entering the atmosphere.

Livestock

Animals used as labour or raised for meat, eggs, dairy, wool, fur, and other products.

Methane

A greenhouse gas that traps heat in the atmosphere. It is produced by enteric fermentation, the burning of fossil fuels, and the decomposing of organic matter.

Nitrous oxide

A greenhouse gas that traps heat in the atmosphere. The rise in nitrous oxide is mainly due to the use of fertilizer.

Nuclear energy

The energy that comes from the splitting of atoms. Nuclear energy can be used as an alternative to burning fossil fuels, but the waste it produces is highly toxic for many years.

Organic

Something that is made up of living matter.

Per capita

Relating to individual people of a population.

Permafrost

Frozen ground found at or below the surface in polar regions.

Pollutant

A substance that pollutes the atmosphere or water.

Pollution

The introduction of harmful materials into the environment.

Population

The number of people living in a specific area.

Recycle

To use something again or make it into something new.

Renewable energy

An energy source that can be used again and again, in contrast to one that eventually gets used up. Examples include solar, wind, and hydroelectric power. Also known as clean energy or green energy.

Rural

Of or relating to the countryside.

Sea level

The level of the sea's surface in relation to other geographic features such as coasts. Rising sea levels are a major consequence of climate change.

Soil erosion

The wearing away of the top layer of soil by natural processes such as wind, rain, and animal activity. Extreme weather due to climate change is leading to increased soil erosion in some areas.

Solar power

Energy from sunlight that is converted into electricity.

Subsistence farming

Raising just enough crops or animals to feed one's family.

Sustainability

The act of using resources in a way that will prevent them from running out or becoming difficult for future generations to find.

Toxic

A substance that is dangerous or deadly to people, animals, or plants.

UNFCCC

United Nations Framework Convention on Climate Change.

Urban

Of or relating to a city.

Wind power

Energy from the wind that is converted into electricity.

Index

Acknowledgements

Dorling Kindersley would like to thank: Georgina Palffy, Jenny Sich, Anna Streiffert-Limerick, and Selina Wood for contributing text, Kelsie Besaw for editorial assistance, Victoria Pyke for proofreading, Elizabeth Wise for indexing, Tanya Mehrotra for jacket design, Rakesh Kumar for DTP design, Priyanka Sharma for jacket editorial coordination and Saloni Singh as managing jackets editor.

The publisher would like to thank the following for their kind permission to reproduce their photographs:
(Key: a-above; b-below/bottom; c-centre; f-far; l-left; r-right; t-top)

4 iStockphoto.com: oatawa (l, tr/Background, crb/Background). **5 iStockphoto.com:** oatawa (cla/Background, tr/Background). **6-7 ESA:** © ESA / NASA. **8-9 iStockphoto.com:** oatawa (Background). **10-11 iStockphoto.com:** oatawa (Background). **20-21 iStockphoto.com:** oatawa (Background). **22-23 iStockphoto.com:** oatawa (Background). **28 Getty Images:** Hindustan Times (l). **29 Getty Images:** Hindustan Times (r). **30-31 TurboSquid:** Devil3d / Dorling Kindersley (cows). **33 123RF.com:** Richard Whitcombe (bl). **34-35 Greenpeace:** Ulet Ifansasti. **38-39 Getty Images:** Michael Markieta / Moment. **41 Getty Images:** Mariama Darame / AFP (br). **42-43 iStockphoto.com:** oatawa (Background). **44-45 iStockphoto.com:** oatawa (Background). **47 Alamy Stock Photo:** Thomas Faull (cb). **49 ESA:** NASA / A. Gerst (cb). **51 Alamy Stock Photo:** Eyal Bartov (cr); PhotoStock-Israel / Eyal Bartov (br). **52-53 DMI:** Steffen M. Olsen. **55 Alamy Stock Photo:** Joao Ponces (clb). **Getty Images:** Alexis Rosenfeld (cb). **Science Photo Library:** NOAA (crb). **56 Ritzau Scanpix:** Jakob Dall / EPA (crb). **57 Getty Images:** Narinder Nanu / AFP (cb). **58 Alamy Stock Photo:** Outback Australia (tl). **Getty Images:** Brook Mitchell (tc). **Paul Sudmals:** (tr). **62-63 Getty Images:** David Gray / Stringer. **64 Getty Images:** Gabriel Bouys / AFP (bc); Jonas Gratzer / LightRocket (br). **65 Alamy Stock Photo:** FEMA (bc); Brais Seara (br). **Getty Images:** Orlando Sierra / AFP (bl). **Panos Pictures:** G.M.B. Akash (fbr). **66-67 iStockphoto.com:** oatawa (Background). **68-69 iStockphoto.com:** oatawa (Background). **70 Alamy Stock Photo:** COP21 (br). **72 Getty Images:** VCG (bl). **73 Alamy Stock Photo:** Steve Morgan (bl). **74-75 Alamy Stock Photo:** Aerial-Photos.com. **77 Alamy Stock Photo:** Rob Arnold (cra). **78 Alamy Stock Photo:** BrazilPhotos / J.R.Ripper (cla); Andrew Pearson (clb); Johner Images (bl). **Getty Images:** Cavan Images (tl). **79 Alamy Stock Photo:** Sandipan Dutta (cra); Manit Larpluechai (crb). **Dreamstime.com:** Shannon Fagan **(br). SuperStock:** Universal Images (tr). **80 Getty Images:** Paulo Fridman / Bloomberg (cb). **81 Alamy Stock Photo:** Oreolife (tr). **82 Getty Images:** Bloomberg / Camilla Cerea (cb). **TurboSquid:** 3d_molier International (c); jasenluxchambers (cra). **83 Alamy Stock Photo:** Lena Kuhnt (cra). **TurboSquid:** 3d_molier International (ca). **84 Alamy Stock Photo:** Bernard Golden (bl). **Dreamstime.com:** Key909 (tl/flag, tc/flag). **Getty Images:** Ricky Carioti / The Washington Post (tl); Bas Czerwinski / AFP (tc); Xaume Olleros / Anadolu Agency (cb). **85 Alamy Stock Photo:** Joerg Boethling (cb); Ian Mckenzie (tc). **Dreamstime.com:** Key909 (tl/flag, tc/flag). **Getty Images:** Visual China Group (tl); Edwin Remsberg / VWPics / Universal Images Group (bl). **86-87 Alamy Stock Photo:** Gonzales Photo / Astrid Maria Rasmussen. **88 Alamy Stock Photo:** Per Grunditz (bc). **89 Alamy Stock Photo:** Michael Wheatley Photography (clb). **90-91 iStockphoto.com:** oatawa (Background)

OTHER REFERENCES

16-17 International Civil Aviation Organization (ICAO): ICAO has coordinated a global strategy to address carbon emissions from aviation. **24-25 NASA:** SEDAC, IFPRI, WCMC, The World Bank, and CIAT. **26 Global Carbon Atlas / The Global Carbon Project:** www.globalcarbonatlas.org / Our World in Data | https://ourworldindata.org/ (bl). **26-27 Climate Watch. 2018. Washington, DC: World Resources Institute:** CAIT Climate Data Explorer. 2019. Country Greenhouse Gas Emissions. Washington, DC: World Resources Institute. **30-31 Food and Agriculture Organization of the United Nations (FAO). 31 Food and Agriculture Organization of the United Nations. 32-33 Global Forest Watch:** GLAD Alerts Footprint. **36-37 IEA:** IEA (2019). CO2 Emissions from Fuel Combustion, Highlights. https://webstore.iea.org/co2-emissions-from-fuel-combustion-2019-highlights. All rights reserved. As modified by Dorling Kindersley. **38 International Civil Aviation Organization (ICAO):** ICAO has coordinated a global strategy to address carbon emissions from aviation. (clb). **46-47 NASA:** Scientific Visualization Studio / Data provided by Robert B. Schmunk (NASA / GSFC GISS). **50-51 National Snow and Ice Data Center / NSIDC:** Sea Ice Index. **54-55 © 2020 C40 Cities Climate Leadership Group, Inc. All rights reserved.:** (map). **OHC by IAP:** Cheng L. *, K. Trenberth, J. Fasullo, T. Boyer, J. Abraham, J. Zhu, 2017: Improved estimates of ocean heat content from 1960 to 2015, Science Advances. 3,e1601545c. **56-57 © KNMI. 58-59 Western Australian Land Information Authority (Landgate):** MyFireWatch. **60-61 The Royal Society:** Tim Newbold © 2018 The Authors. Published by the Royal Society under the terms of the CC BY 4.0. **70-71 © 2019 The World Bank Group:** World Development Indicators, The World Bank / Our World in Data | https://ourworldindata.org/. **Copyright 2020 Climate Action Tracker:** Copyright © 2020 by Climate Analytics and NewClimate Institute. All rights reserved. (stars). **72-73 © IRENA 2019:** Renewable capacity statistics 2019, International Renewable Energy Agency. **76-77 © GWEC – Global Wind Energy Council:** Global Wind Report 2019. **78-79 Atlas of Forest Landscape Restoration Opportunities:** GPFLR / WRI. **82-83 Joseph Poore | Dr. Thomas Nemecek:** Data from Poore & Nemecek SCIENCE 360:987 (2018) (GHG data)

All other images © Dorling Kindersley
For further information see: www.dkimages.com

ARCTIC OCEAN

Chukchi Sea

Beaufort Sea

Queen Elizabeth Islands

Ellesmere Island

Greenland

Greenland Sea

Bering Strait

Brooks Range

Yukon

Mackenzie

Victoria Island

Great Bear Lake

Baffin Island

Baffin Bay

Denmark Strait

Norwegian Sea

△ Denali (Mount McKinley) 6,194 m (20,320 ft)

Great Slave Lake

Hudson Bay

Davis Strait

Labrador Sea

Iceland

Bering Sea

Aleutian Basin

Coast Mountains

Canadian Shield

North America

Laurentian Mountains

British Isles

North Sea

Aleutian Islands

Aleutian Trench

Gulf of Alaska

Rocky Mountains

Great Lakes

Vancouver Island

Great Plains

Missouri

Appalachian Mts

North American Basin

Iberian Peninsula

Me

Mendocino Fracture Zone

Mississippi

Azores

Atlas Mountains

Murray Fracture Zone

Sierra Nevada

Sierra Madre Occidental

Sierra Madre Oriental

Gulf of Mexico

West Indies

Madeira

Canary Islands

S a h a

Ahag

Hawaiian Islands

Hawaii

Clarion Fracture Zone

Yucatán Peninsula

Greater Antilles

Cape Verde Islands

S a h

Line Islands

Kiritimati

Clipperton Fracture Zone

Middle America Trench

Caribbean Sea

Lesser Antilles

Mid-Atlantic Ridge

ATLANTIC

Niger

Polynesia

PACIFIC

Galápagos Islands

Orinoco

Guiana Highlands

OCEAN

Gulf of Guinea

Marquesas Islands

OCEAN

Peru-Chile Trench

Amazon

Amazon Basin

SOUTH AMERICA

Brazil Basin

Tuamotu Islands

Peru Basin

Nazca Ridge

Andes

Planalto de Mato Grosso

Brazilian Highlands

Brazil Basin

Angol

Basin

Pitcairn Island

Sala y Gomez Ridge

Gran Chaco

Río Grande Rise

Tubuai Islands

Easter Island

Roggeveen Basin

Aconcagua △ 6,959 m (22,837 ft)

Paraná

Mid-Atlantic Ridge

C

Southwest Pacific Basin

East Pacific Rise

Eltanin Fracture Zone

Andes

Pampas

Argentine Basin

Patagonia

Falkland Islands

America-Antarctic Ridge

Tierra del Fuego

South Georgia

S

Cape Horn

Scotia Sea

KEY

△ mountain

— river

Southeast Pacific Basin

Drake Passage

Antarctic Peninsula

Bellingshausen Sea

Weddell Plain

Weddell Sea